Developing Num

MEASURES, SHAPE AND SPACE

ACTIVITIES FOR THE DAILY MATHS LESSON

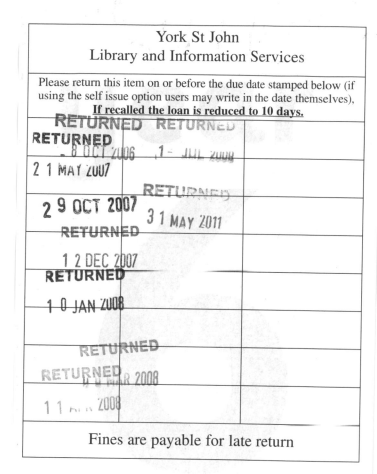
Dave Kirkby

A & C BLACK

Contents

Reprinted 2002, 2006

Published 2001 by A & C Black Publishers Limited
38 Soho Square, London W1D 3HB

ISBN-10: 0-7136-5885-1
ISBN-13: 978-0-7136-5885-9

Copyright text © Dave Kirkby, 2001
Copyright illustrations © Gaynor Berry, 2001
Copyright cover illustration © Charlotte Hard, 2001
Editors: Lynne Williamson and Marie Lister

The author and publishers would like to thank Madeleine Madden and Corinne McCrum for their advice in producing this series of books.

A CIP catalogue record for this book is available from the British Library.

A & C Black uses paper produced with elemental chlorine-free pulp, harvested from managed sustainable forests.

Printed in Great Britain by Caligraving Ltd, Thetford, Norfolk.

Introduction

Developing Numeracy: Measures, Shape and Space is a series of seven photocopiable activity books designed to be used during the daily maths lesson. They focus on the fourth strand of the National Numeracy Strategy *Framework for teaching mathematics*. The activities are intended to be used in the time allocated to pupil activities; they aim to reinforce the knowledge, understanding and skills taught during the main part of the lesson and to provide practice and consolidation of the objectives contained in the framework document.

Year 6 supports the teaching of mathematics by providing a series of activities which develop essential skills in measuring, and exploring pattern, shape and space. On the whole the activities are designed for children to work on independently, although this is not always possible and occasionally some children may need support.

Year 6 encourages children to:

- convert smaller units of measure to larger units and vice versa;
- know imperial units;
- record estimates and readings from scales to a suitable degree of accuracy;
- calculate area and perimeter of simple compound shapes;
- describe and visualise 3-D shapes;
- classify quadrilaterals;
- recognise where a shape will be after reflections, rotations and translations;
- read and plot co-ordinates in all four quadrants;
- use a protractor to measure and draw angles to the nearest degree;
- calculate angles in a triangle and around a point.

Extension

Many of the activity sheets end with a challenge (**Now try this!**) which reinforces and extends the children's learning, and provides the teacher with the opportunity for assessment. On occasion, you may wish to read out the instructions and explain the activity before the children begin working on it. The children may need to record their answers on a separate piece of paper.

Organisation

Very little equipment is needed, but it will be useful to have available rulers, scissors, coloured pencils, counters, dice, squared paper, geoboards, interlocking cubes, 2-D and 3-D shapes, protractors, and small mirrors. You will need to provide stopwatches for page 17 and 5-minute timers for page 58.

The children should also have access to measuring equipment to give them practical experience of length, mass and capacity.

To help teachers to select appropriate learning experiences for the children, the activities are grouped into sections within each book. However, the activities are not expected to be used in that order unless otherwise stated. The sheets are intended to support, rather than direct, the teacher's planning.

Some activities can be made easier or more challenging by masking and substituting some of the numbers. You may wish to re-use some pages by copying them onto card and laminating them, or by enlarging them onto A3 paper.

Teachers' notes

Very brief notes are provided at the foot of each page giving ideas and suggestions for maximising the effectiveness of the activity sheets. These can be masked before copying.

Structure of the daily maths lesson

The recommended structure of the daily maths lesson for Key Stage 2 is as follows:

Start to lesson, oral work, mental calculation	5–10 minutes
Main teaching and pupil activities *(the activities in the **Developing Numeracy** books are designed to be carried out in the time allocated to pupil activities)*	about 40 minutes
Plenary *(whole-class review and consolidation)*	about 10 minutes

Whole-class warm-up activities

The following activities provide some practical ideas which can be used to introduce or reinforce the main teaching part of the lesson.

Measures

Comparison activities

Ask the children to estimate the order of a set of objects based on a given measure. For example, provide a set of five different objects labelled A to E. The children, in pairs, decide which is the lightest, the next lightest, and so on up to the heaviest. The objects are weighed (in a chosen unit) and the true order determined. Compare this with the estimates.

Conversion activities

Practise conversion between metric and imperial units, for example, between pints and litres. Make a set of cards which show capacities in litres (5 litres, 10 litres, 1 litre, and so on). Shuffle the cards and hold them up, one at a time. The children show the equivalent number of pints on numbered cards or fans. Discuss their answers at each stage.

24-hour clock game

Provide each child with a clock face with rotating hands. (Alternatively, they could use a clock face drawn on a sheet of paper and two pointers for the hands, or a paper plate with a clock face drawn on and pointers attached with a split pin.) Show the children a 24-hour time, for example 21:35. Ask the children to show the time as a 12-hour time on their clock faces.

Shape and space

Shape guess

Make a 4 x 4 grid numbered 1 to 16. Draw different shapes in the cells. Photocopy the grid and give one copy to each pair of children, along with a set of counters. Choose a shape and describe it in stages, for example: *It is a quadrilateral.* The children eliminate shapes on their grid (i.e. those which are not quadrilaterals) by placing counters on them. Give the next stage of the description, for example: *It has a right angle.* Continue until there is only one shape left uncovered. Ask the children the number of the shape.

Name the shape 'catapult'

This is played like the traditional game of 'Hangman'. Draw on the board a picture of the class sitting on a giant catapult. Draw five ropes holding the catapult down. Explain that the word to guess is the name of a shape. The children guess a letter, for example 'a'. If it appears in the word, write it in. If not, rub out one of the ropes. The aim of the game is to guess the name of the shape before all five ropes are cut, releasing the catapult and propelling the whole class into the air.

Class co-ordinates

Seat the children in neat rows and columns, for example, 30 children in 6 rows and 5 columns. Give each column a horizontal co-ordinate (–2 to 2), and each row a vertical co-ordinate (–3 to 2). Play 'Stand-up games', by giving instructions such as: *Stand up if your horizontal co-ordinate is -1. Stand up if your vertical co-ordinate is less than 1.* The children should sit down again between instructions.

Paper plate angles

Use two different-coloured circles of paper, identical in size (you could use two paper plates). Cut each circle with a straight line from the edge to the centre. Intersect the circles so that you can rotate one of them to demonstrate an angle. Show an angle to the children and ask them to estimate its size in degrees. Choose a child to measure the angle using a protractor and see how close the estimates are. Repeat for different acute, obtuse and reflex angles.

Signposts

The map is in ⟨kilometres⟩.
The sign is in ⟨miles⟩.

8 km is approximately 5 miles.

• **Fill in the missing distances.**

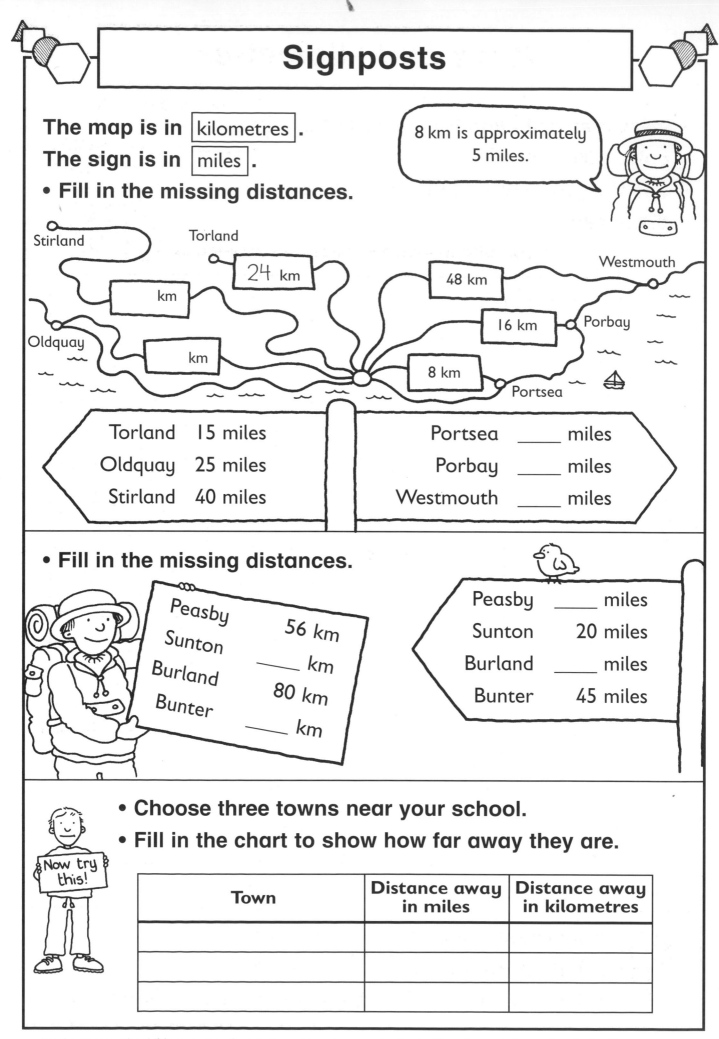

Stirland

Torland

24 km

____ km

48 km

Westmouth

16 km

Porbay

____ km

Oldquay

____ km

8 km

Portsea

Torland	15 miles
Oldquay	25 miles
Stirland	40 miles

Portsea	____ miles
Porbay	____ miles
Westmouth	____ miles

• **Fill in the missing distances.**

Peasby	56 km
Sunton	____ km
Burland	80 km
Bunter	____ km

Peasby	____ miles
Sunton	20 miles
Burland	____ miles
Bunter	45 miles

• **Choose three towns near your school.**
• **Fill in the chart to show how far away they are.**

Now try this!

Town	Distance away in miles	Distance away in kilometres

Teachers' note The children may need reminding of how to convert miles to kilometres and vice versa. For the extension activity, provide mileage charts for the children to refer to. Alternatively, write the towns and the distances in either miles or kilometres on the chart before photocopying.

Developing Numeracy
Measures, Shape and Space
Year 6
© A & C Black

6

Miles and kilometres

- **Draw a straight line from** (0, 0) **to the point** (100, 160).

Remember, 100 miles ⌒ 160 km.

- **Use the graph to find the approximate distances.**

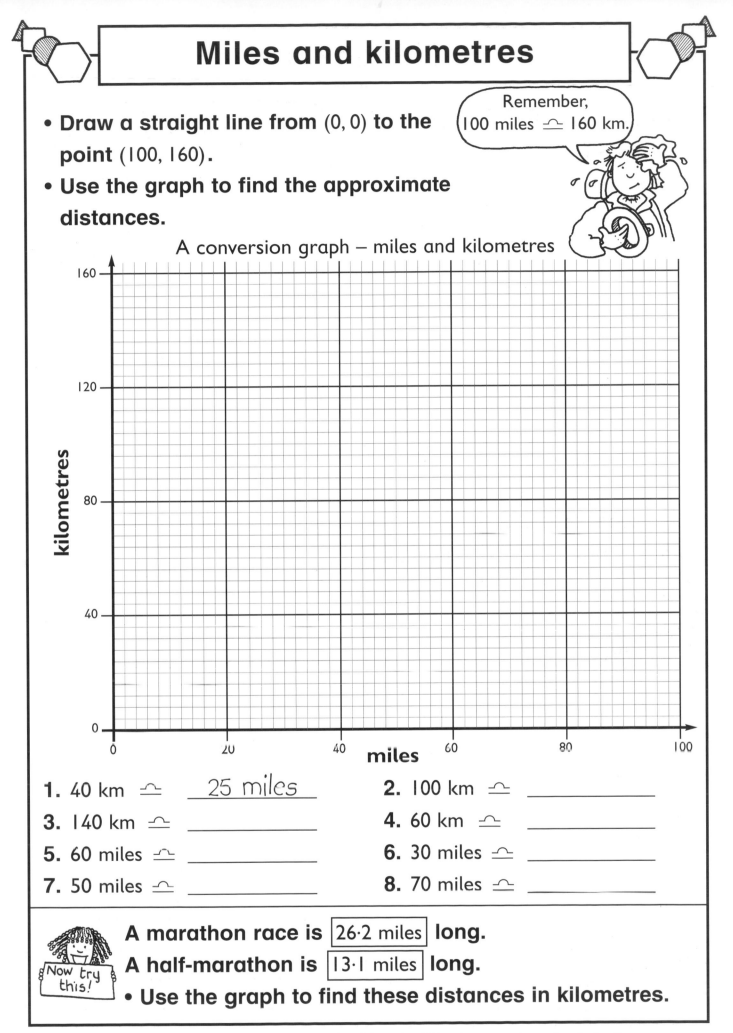

A conversion graph – miles and kilometres

(vertical axis: kilometres — 0, 40, 80, 120, 160)
(horizontal axis: miles — 0, 20, 40, 60, 80, 100)

1. 40 km ⌒ _25 miles_

2. 100 km ⌒ _____

3. 140 km ⌒ _____

4. 60 km ⌒ _____

5. 60 miles ⌒ _____

6. 30 miles ⌒ _____

7. 50 miles ⌒ _____

8. 70 miles ⌒ _____

Now try this!

A marathon race is | 26·2 miles | **long.**

A half-marathon is | 13·1 miles | **long.**

- **Use the graph to find these distances in kilometres.**

Teachers' note Revise the 'approximately equals' sign and explain to the children how to read a conversion graph. For example: to convert from miles to kilometres, read the mileage on the horizontal axis, draw a vertical line from here to the conversion line, then draw a horizontal line from this point to the vertical axis. Read the value in kilometres.

**Developing Numeracy
Measures, Shape and Space
Year 6**
© A & C Black

Walk this way!

- **Follow the path. Convert to** metres **or** kilometres **as you go.**

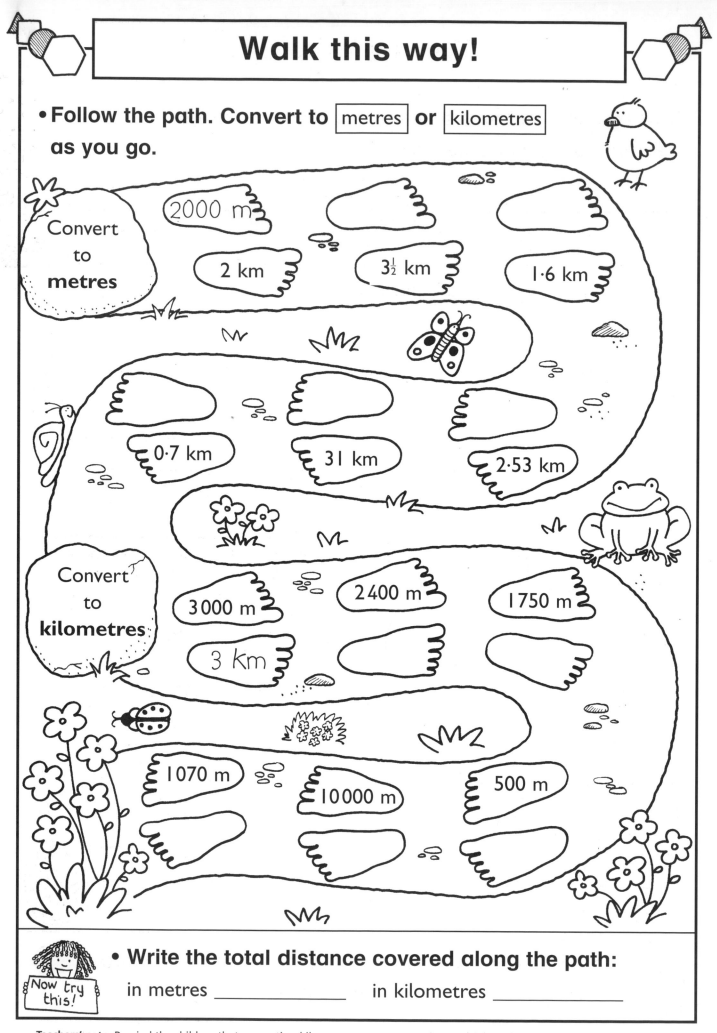

Convert to **metres**

2000 m

2 km   3½ km   1·6 km

0·7 km   31 km   2·53 km

Convert to **kilometres**

3000 m   2400 m   1750 m

3 km

1070 m   10000 m   500 m

Now try this!

- **Write the total distance covered along the path:**

   in metres _____   in kilometres _____

Teachers' note Remind the children that converting kilometres to metres requires multiplying by 1000 (thus sliding the digits three places to the left), and converting metres to kilometres requires dividing by 1000 (thus sliding the digits three places to the right). As an extension, mask the values and ask the children to write in one value for each pair of feet for a partner.

Developing Numeracy
Measures, Shape and Space
Year 6
© A & C Black

Tennis mix-up

This tennis court was marked out using a yard stick.

- Read the measurements in [feet].

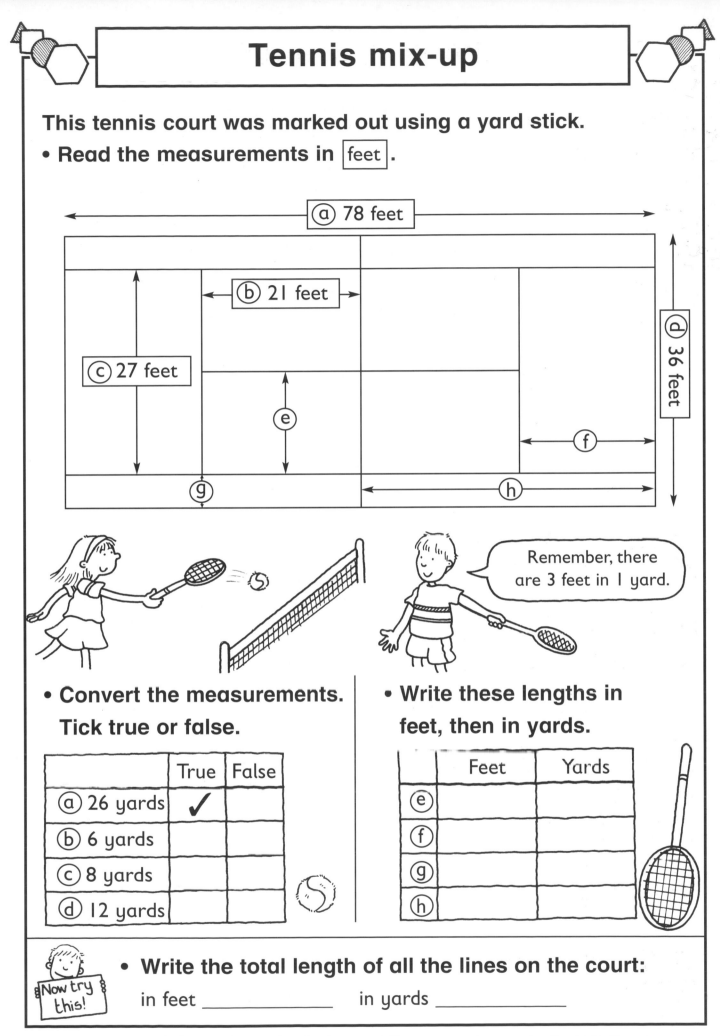

(a) 78 feet

(b) 21 feet

(c) 27 feet

(d) 36 feet

e

f

g

h

Remember, there are 3 feet in 1 yard.

- **Convert the measurements. Tick true or false.**

	True	False
(a) 26 yards	✓	
(b) 6 yards		
(c) 8 yards		
(d) 12 yards		

- **Write these lengths in feet, then in yards.**

	Feet	Yards
(e)		
(f)		
(g)		
(h)		

Now try this!

- **Write the total length of all the lines on the court:**

  in feet _____   in yards _____

Teachers' note Revise the fact that there are three feet in one yard. The measurements are exact for a standard tennis court. Explain to the children that a tennis court is symmetrical and they can work out the missing lengths from the other lengths.

Developing Numeracy
Measures, Shape and Space
Year 6
© A & C Black

Bubble game

- **Play with a partner. You need 16 counters.**

☆ Cover each bubble with a counter.
☆ Take turns to remove a counter. Say the distance in **metres**.
☆ If your partner agrees, keep the counter. Otherwise, replace it.
☆ The winner is the player with the most counters.

For 3500 mm, say 'three point five metres'.

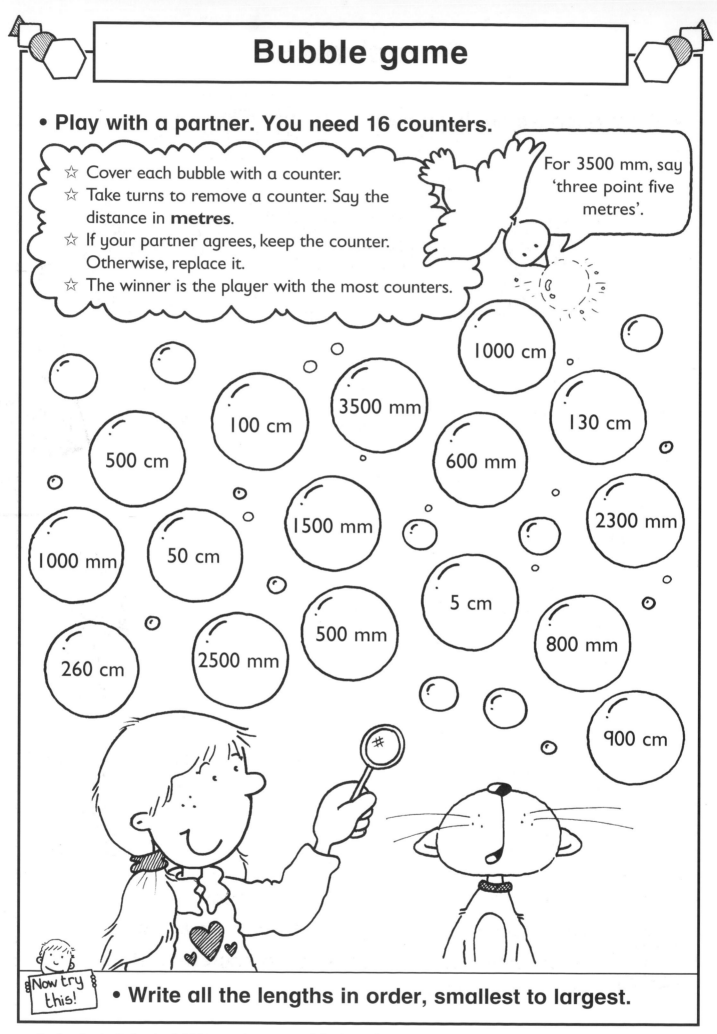

1000 cm

3500 mm

100 cm

130 cm

500 cm

600 mm

1500 mm

2300 mm

1000 mm

50 cm

5 cm

500 mm

800 mm

260 cm

2500 mm

900 cm

Now try this!

- **Write all the lengths in order, smallest to largest.**

Teachers' note Each pair of children needs one copy of the sheet. If necessary, revise converting centimetres and millimetres to metres. The children should agree on the correct answers together. Any they are not sure of should be noted and discussed during the plenary session.

Developing Numeracy
Measures, Shape and Space
Year 6
© A & C Black

Parcel polka

- **Write each weight in** grams .

> Remember,
> 1000 g = 1 kg.

1.

1½ kg

1500 g

2.

5 kg

_____ g

3.

2·7 kg

_____ g

4.

6·65 kg

_____ g

5.

3·525 kg

_____ g

6.

0·55 kg

_____ g

- **Write each weight in** kilograms .

7.

2000 g

2 kg

8.

1600 g

_____ kg

9.

3750 g

_____ kg

10.

600 g

_____ kg

11.

80 g

_____ kg

12.

1050 g

_____ kg

A 1 kilogram parcel costs 70p **to post.**
- **Calculate the cost of posting each parcel in questions 1 to 6. First round up each weight to the nearest kilogram.**

Now try this!

Teachers' note Remind the children that converting kilograms to grams requires multiplying by 1000 (thus sliding the digits three places to the left), and converting grams to kilograms requires dividing by 1000 (thus sliding the digits three places to the right).

**Developing Numeracy
Measures, Shape and Space
Year 6**
© A & C Black

11

Gold panning

Charlie and Billy August were gold panners for ten years.
They recorded their finds.

30 g ⌒ 1 oz

- Write the equivalent values.

Gold finds 1520 to 1530

1520	__1__	oz	30 g	1526	_____	60 g
1521	_____	oz	45 g	1527	_____	90 g
1522	_____		15 g	1528	_____	135 g
1523	_____		150 g	1529	_____	105 g
1524	_____		120 g	1530	_____	165 g
1525	_____		75 g	Total	_____	_____

Charlie August

Gold finds 1520 to 1530

1520	4 oz	120 g	1526	6 oz	_____
1521	9 oz	_____	1527	3 oz	_____
1522	2 oz	_____	1528	12 oz	_____
1523	10 oz	_____	1529	$1\frac{1}{4}$ lb	_____
1524	5 oz	_____	1530	2 lb	_____
1525	1 oz	_____	Total	_____	_____

Billy August

Now try this!

Each ounce was worth about $£3$.

- **How much money did each man make in 10 years?**

Charlie _____   Billy _____

Teachers' note Remind the children that there are 16 ounces in one pound and ensure they are familiar with the notations 'oz' and 'lb'. As a further extension, the children could calculate approximately how much a gram of gold was worth.

Developing Numeracy
Measures, Shape and Space
Year 6
© A & C Black

Pet paradise

The scales show the amount of rabbit food sold over two weeks.

- **Write each amount in** pounds.

1 kg ⌒ 2 lb

Week 1

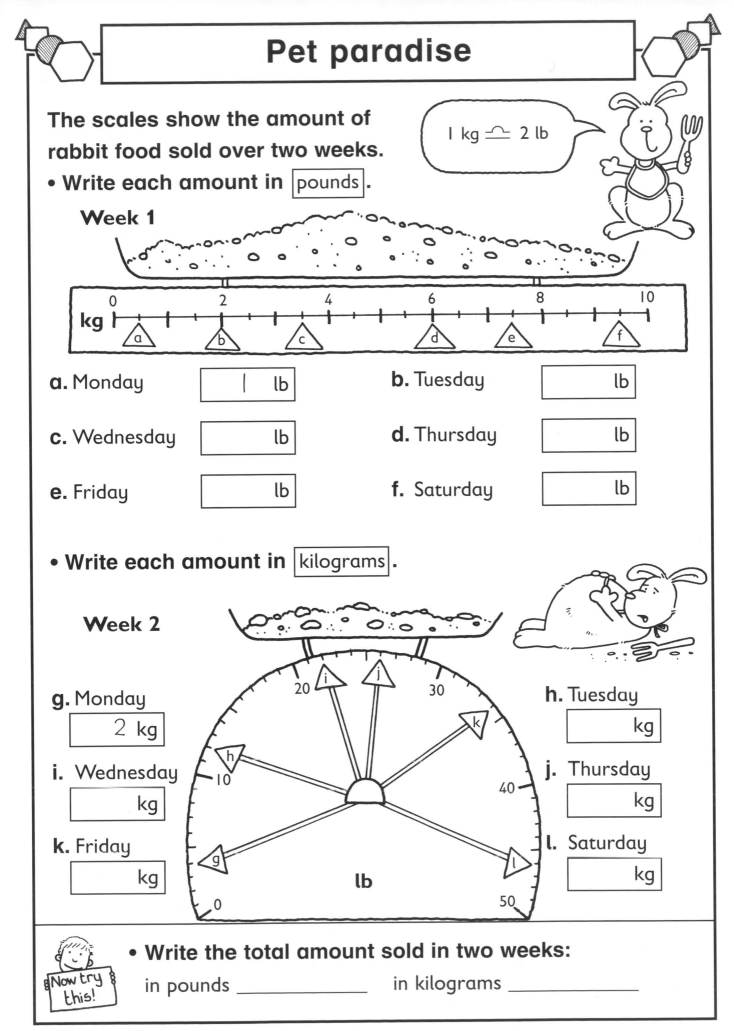

a. Monday	1 lb	**b.** Tuesday	lb
c. Wednesday	lb	**d.** Thursday	lb
e. Friday	lb	**f.** Saturday	lb

- **Write each amount in** kilograms.

Week 2

g. Monday	2 kg	**h.** Tuesday	kg
i. Wednesday	kg	**j.** Thursday	kg
k. Friday	kg	**l.** Saturday	kg

Now try this!

- **Write the total amount sold in two weeks:**

  in pounds _____   in kilograms _____

Teachers' note Ensure the children are familiar with the 'approximately equals' sign. As a further extension, the children could repeat the activity using the more accurate equivalence 1 kg = 2.2 lb.

Developing Numeracy
Measures, Shape and Space
Year 6
© A & C Black

Capacity match game

• **Work with a partner.**

☆ Cut out the cards. Spread them face down.

☆ Take turns to pick a card. Place it face up in front of you.

☆ When all the cards have been collected, score points for matching cards:

      a matching pair scores 2 points

      a matching trio scores 5 points

☆ The winner is the player with the most points.

Remember, there are 1000 ml in 1 litre.

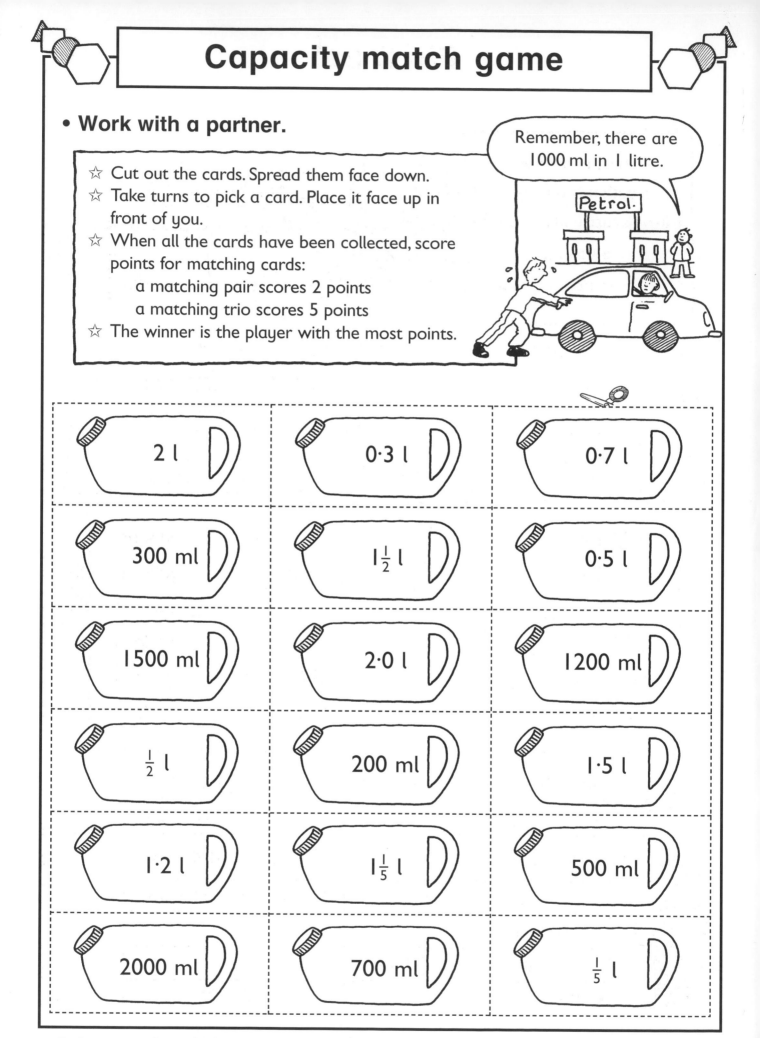

Teachers' note Each pair of children needs one copy of the sheet. Remind the children that there are 1000 ml in 1 litre and ensure they are familiar with the notation 'ml'. As an extension activity, the children can pick pairs of cards and find the difference in millilitres.

**Developing Numeracy
Measures, Shape and Space
Year 6**
© A & C Black

14

Fill it up!

Petrol readings

1 gallon is 8 pints

1 litre is just under 2 pints

4·5 litres is

approximately 1 gallon

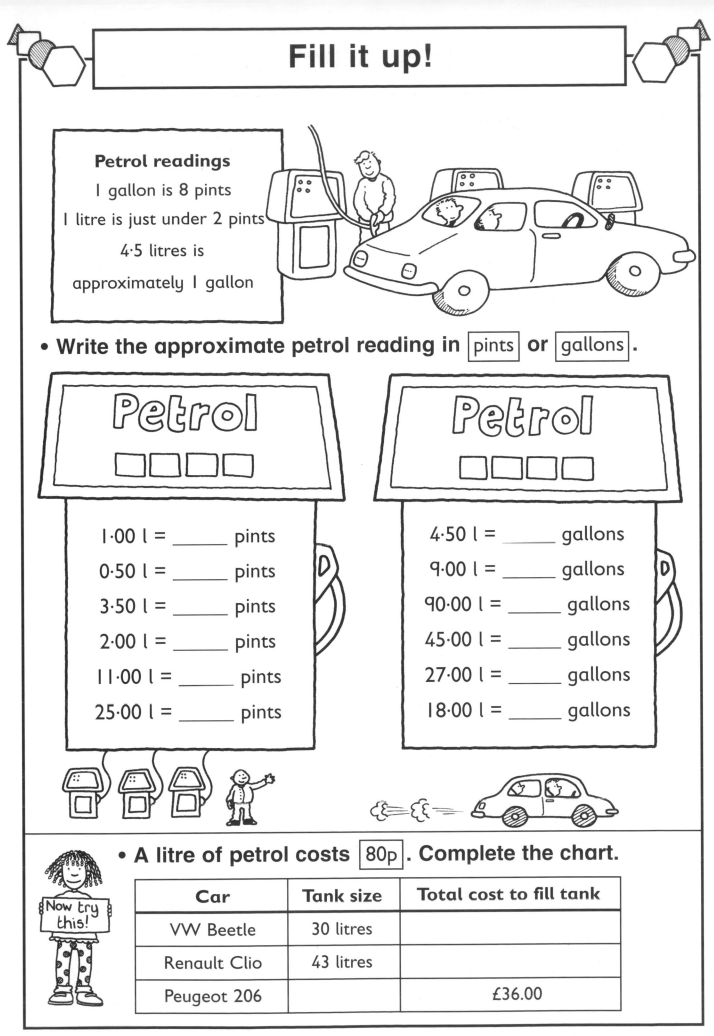

• **Write the approximate petrol reading in** pints **or** gallons .

Petrol

1·00 l = _____ pints

0·50 l = _____ pints

3·50 l = _____ pints

2·00 l = _____ pints

11·00 l = _____ pints

25·00 l = _____ pints

Petrol

4·50 l = _____ gallons

9·00 l = _____ gallons

90·00 l = _____ gallons

45·00 l = _____ gallons

27·00 l = _____ gallons

18·00 l = _____ gallons

Now try this!

• **A litre of petrol costs** 80p . **Complete the chart.**

Car	Tank size	Total cost to fill tank
VW Beetle	30 litres	
Renault Clio	43 litres	
Peugeot 206		£36.00

Teachers' note The children could research petrol and diesel prices, tank sizes of different
vehicles and how much it costs to run them.

Developing Numeracy
Measures, Shape and Space
Year 6
© A & C Black

How much?

- **Write each amount in** $\boxed{\text{millilitres}}$.

Remember, 1000 ml = 1 litre.

1. _____ ml

2. _____ ml

3. _____ ml

- **Write each amount in** $\boxed{\text{centilitres}}$.

Remember, 100 cl = 1 litre.

4. _____ cl

5. _____ cl

6. _____ cl

- **Write each amount in** $\boxed{\text{litres}}$.

7. 220 cl _____ l

8. 320 ml _____ l

9. 450 cl _____ l

10. 35 ml _____ l

11. 150 ml _____ l

12. 75 cl _____ l

Now try this!

- **Write all the amounts in** $\boxed{\text{centilitres}}$.

Teachers' note Revise the fact that there are 1000 ml or 100 cl in 1 litre and ensure the children are familiar with the notations 'ml' and 'cl'.

Developing Numeracy
Measures, Shape and Space
Year 6
© A & C Black

Seconds out!

• **You need a stopwatch.**

☆ Look at the time on the first chart.
☆ Start the stopwatch (but don't look at it). Stop the watch when you think the right amount of time has passed.
☆ Read the stopwatch and write the time. This is your estimate.
☆ Find the **difference** in seconds between the actual time and your estimate. This is your score.
☆ A low score is a good score!

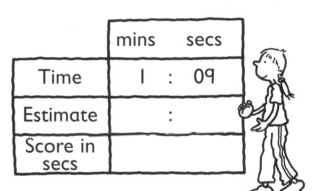

	mins	secs
Time	0 :	20
Estimate	:	
Score in secs		

	mins	secs
Time	0 :	45
Estimate	:	
Score in secs		

	mins	secs
Time	2 :	17
Estimate	:	
Score in secs		

	mins	secs
Time	1 :	09
Estimate	:	
Score in secs		

	mins	secs
Time	1 :	44
Estimate	:	
Score in secs		

	mins	secs
Time	1 :	39
Estimate	:	
Score in secs		

Total score

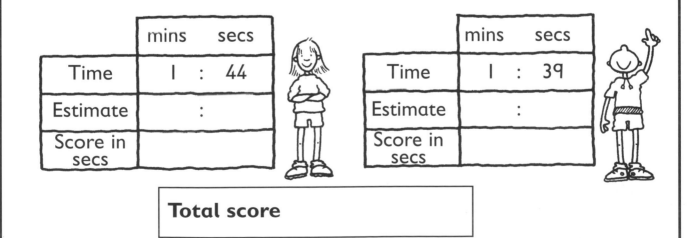

Teachers' note Ask the children to repeat the activity to see if they can improve their score. As an extension activity, the children could work in pairs: one child holds and reads the watch, while the other child estimates the time period. The children could swap roles and compete to see whose estimate is the nearest.

**Developing Numeracy
Measures, Shape and Space
Year 6
© A & C Black**

Anagram fun

- Rearrange the letters to find a unit of measure.
- Write a relationship for each measure.

Anagram	Unit	Relationship
remet	metre	1 m = 100 cm
nopud		
foto		
limirillet		
tretemince		
trile		
loglan		
dray		
timelimrel		
grolimak		
chin		
nittlerice		
marg		
unoce		
klotmeeri		
nipt		
lime		

Now try this!

- Write a list of measuring instruments, such as protractor, balance, ruler, tape measure.
- Make up anagrams of them for a partner to solve.

Teachers' note Some children may need reminding of the prefixes 'cent' and 'kilo' when completing the relationship column. As a further activity, the children could make up anagrams of shape names.

Developing Numeracy
Measures, Shape and Space
Year 6
© A & C Black

Same but different

- Draw five **different** shapes, each with a perimeter of ☐ 10 cm ☐. Use horizontal and vertical lines.

This shape has a perimeter of 10 cm.

- On squared paper, draw eight different shapes. Each shape should have a perimeter of ☐ 12 cm ☐.

Teachers' note Encourage the children to recognise that shapes with a perimeter of 10 cm do not all have the same area (some have an area of 4 cm², some 5 cm², and some 6 cm²).

**Developing Numeracy
Measures, Shape and Space
Year 6
© A & C Black**

Odd ones out

• **Look at these shapes.**

1. Eleven shapes have the same area.

   Which three shapes are different? _____

2. Eleven shapes have the same perimeter.

   Which three shapes are different? _____

3. Which shapes are **not** octagons? _____

Now try this!

• **Which five shapes have** line symmetry **?** _____
• **Draw the lines of symmetry on the shapes.**

Teachers' note You could ask the children to create their own set of shapes like this by drawing on squared paper, using five or six squares for each shape. They can then cut them out and sort them according to their properties.

Developing Numeracy
Measures, Shape and Space
Year 6
© A & C Black

Area and perimeter

- **Work out the area and perimeter of each shape. Fill in the chart. The shapes are not drawn to scale.**

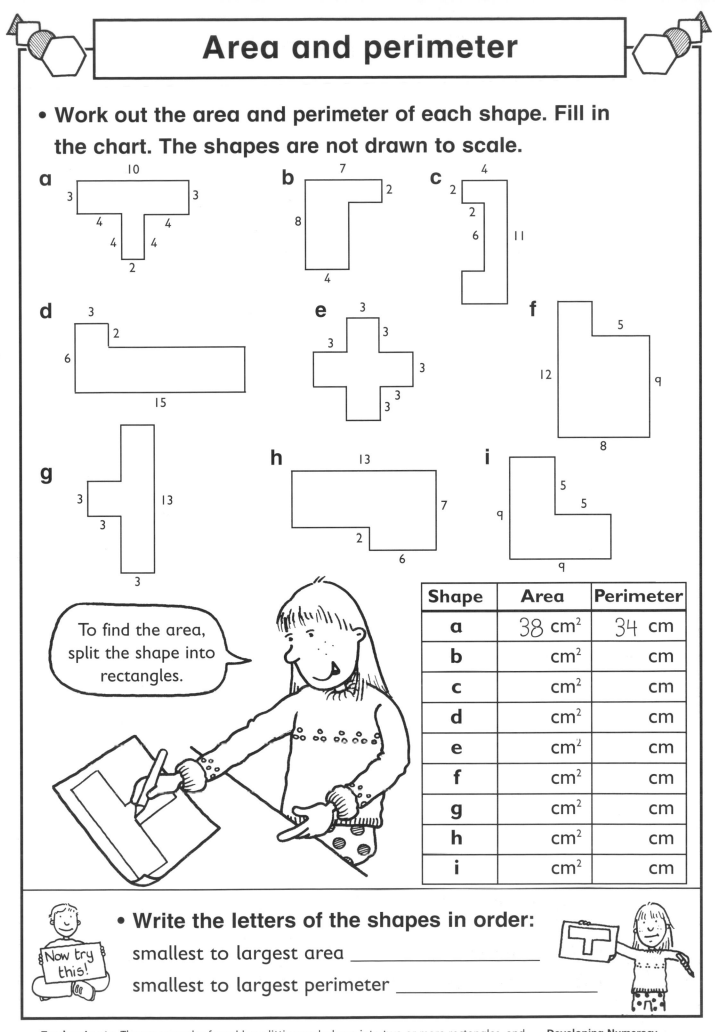

To find the area, split the shape into rectangles.

Shape	Area	Perimeter
a	38 cm²	34 cm
b	cm²	cm
c	cm²	cm
d	cm²	cm
e	cm²	cm
f	cm²	cm
g	cm²	cm
h	cm²	cm
i	cm²	cm

- **Write the letters of the shapes in order:**

Now try this!

smallest to largest area _____

smallest to largest perimeter _____

Teachers' note The areas can be found by splitting each shape into two or more rectangles, and finding the area of each. Alternatively, the areas can be found by considering the area of the 'surrounding' rectangle, and subtracting the areas of pieces cut out from this. Encourage the children to write the lengths of all the missing sides to work out the perimeters.

Developing Numeracy
Measures, Shape and Space
Year 6
© A & C Black

Different squares

- **Join the dots to make** squares .
- **Make each square match the area shown.**

Use a geoboard and a rubber band to help you.

area = 1 unit²

area = 2 units²

area = 4 units²

area = 5 units²

area = 8 units²

area = 9 units²

area = 10 units²

area = 16 units²

Now try this!

- **Draw** rectangles **with different areas on each board.**

Teachers' note The oblique squares are more difficult to find. Their areas are best found by finding the areas of the right-angled triangles **outside** the square, and subtracting these from the area of the whole grid. For the extension activity, the children could draw the rectangles in a different colour.

Developing Numeracy
Measures, Shape and Space
Year 6
© A & C Black

Football pitch

These are the measurements of a football pitch, in yards.

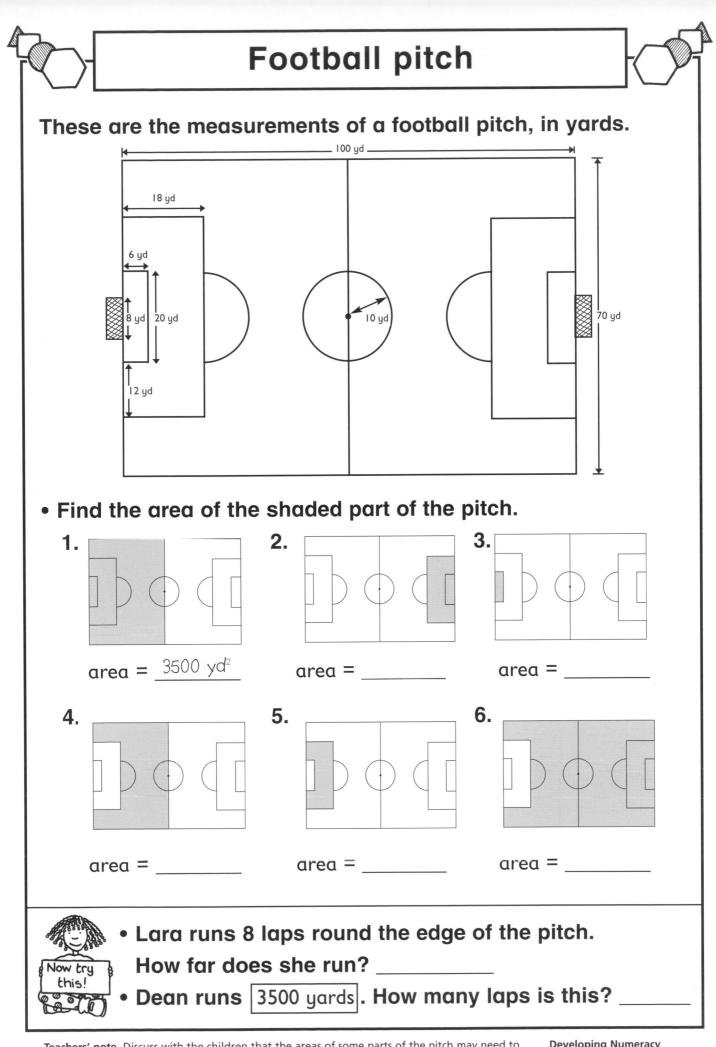

100 yd

18 yd

6 yd

8 yd | 20 yd

12 yd

10 yd

70 yd

• **Find the area of the shaded part of the pitch.**

1.

area = _3500 yd²_

2.

area = _____

3.

area = _____

4.

area = _____

5.

area = _____

6.

area = _____

Now try this!

• **Lara runs 8 laps round the edge of the pitch.**
  How far does she run? _____
• **Dean runs** | 3500 yards | . **How many laps is this?** _____

Teachers' note Discuss with the children that the areas of some parts of the pitch may need to be calculated by subtracting one area from another.

**Developing Numeracy
Measures, Shape and Space
Year 6
© A & C Black**

23

Find the area

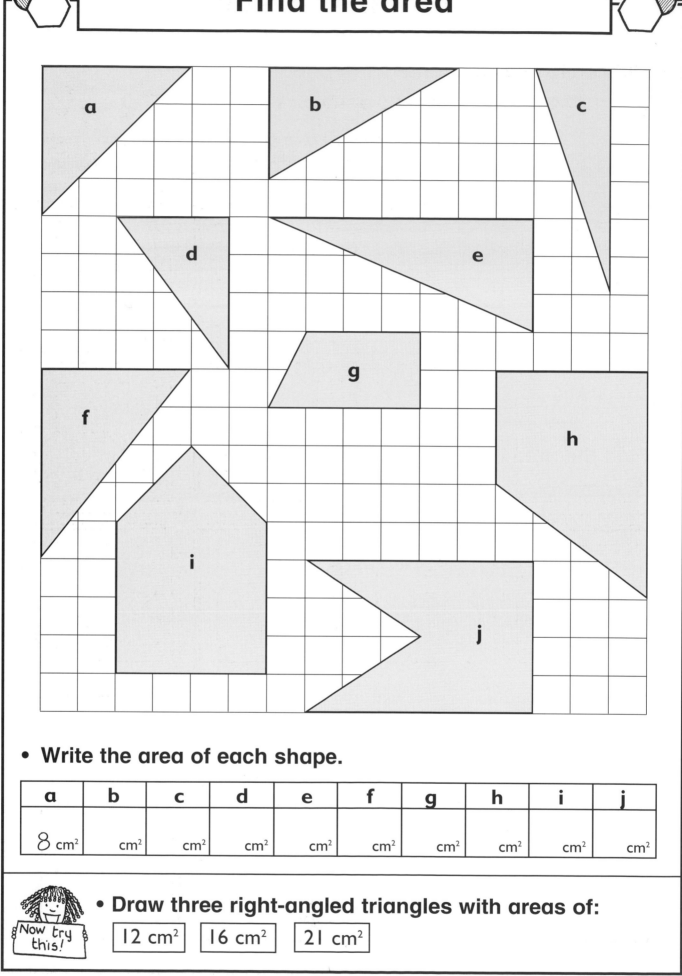

- **Write the area of each shape.**

a	b	c	d	e	f	g	h	i	j
8 cm²	cm²	cm²	cm²	cm²	cm²	cm²	cm²	cm²	cm²

Now try this!

- **Draw three right-angled triangles with areas of:**

| 12 cm² | 16 cm² | 21 cm² |

Teachers' note To find the area of a right-angled triangle, the children should consider the rectangle of which it is a half, then find the area of the rectangle and halve it.

**Developing Numeracy
Measures, Shape and Space
Year 6
© A & C Black**

Matching areas

- **Cut out the cards.**
- **Match the shapes with the same area.**

Work with a partner.

- **Draw different shapes with an area of:**

2 units²   2 units²   3 units²   3 units²

Teachers' note To help children recognise the areas of the shapes, encourage them to break the shapes up into whole squares (area of 1 unit²) and half-squares (area of ½ unit²).

Developing Numeracy
Measures, Shape and Space
Year 6
© A & C Black

Same area, different shape

- **Draw as many different shapes as you can with an area of** $\boxed{6 \text{ cm}^2}$.

My shape has an area of 6 cm².

No, **my** shape has an area of 6 cm²!

- **Write the name of each shape you have drawn.**

Teachers' note One strategy for this exercise is to start with any given shape whose area is 6 cm², then imagine taking a piece off and sticking it somewhere else on the shape.

**Developing Numeracy
Measures, Shape and Space
Year 6
© A & C Black**

Surface areas

- **Build each model using interlocking cubes.**
- **Write the** surface area **of each model (including the base).**

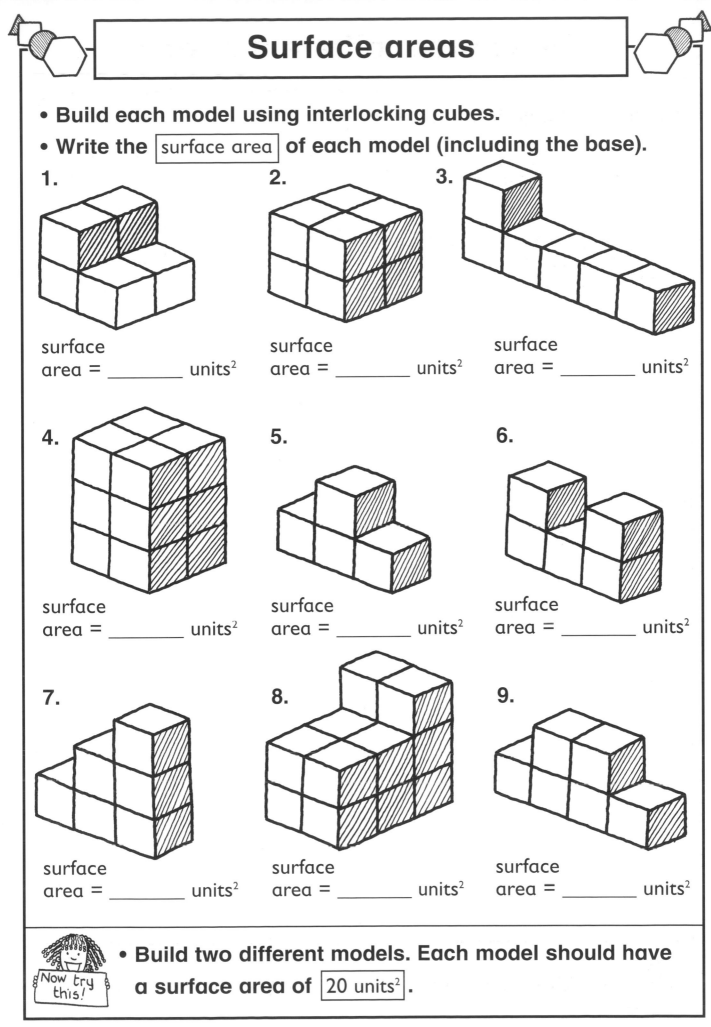

1.

surface
area = _____ units²

2.

surface
area = _____ units²

3.

surface
area = _____ units²

4.

surface
area = _____ units²

5.

surface
area = _____ units²

6.

surface
area = _____ units²

7.

surface
area = _____ units²

8.

surface
area = _____ units²

9.

surface
area = _____ units²

- **Build two different models. Each model should have a surface area of** 20 units² .

Teachers' note Encourage the children to work systematically to find the 'faces', for example:
'top facing', 'bottom facing', 'end facing', 'side facing'.

**Developing Numeracy
Measures, Shape and Space
Year 6
© A & C Black**

All about cubes

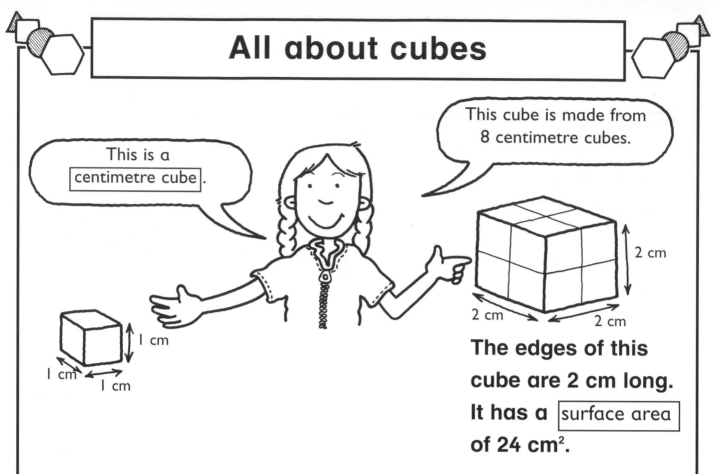

This is a centimetre cube.

This cube is made from 8 centimetre cubes.

The edges of this cube are 2 cm long. It has a surface area of 24 cm².

- Use this information to complete the chart.

Length of edge	Number of cubes needed	Surface area
1 cm		
2 cm	8	24 cm²
3 cm		
4 cm		
5 cm		
6 cm		

Now try this!

- Write the chart entries for a cube whose edges are 10 cm long.

10 cm		

Teachers' note Some children may need to use actual centicubes to help them complete the chart. Discuss the simplest way to find the surface area. Encourage the children to spot a relationship involving square numbers, i.e. for a cube whose side is *n*, the surface area is 6 x *n*².

Developing Numeracy
Measures, Shape and Space
Year 6
© A & C Black

Square cuts

The area of each square is 16 cm^2.

The mid-point of each side is marked.

One has been done for you.

• Write the area of each part inside the square.

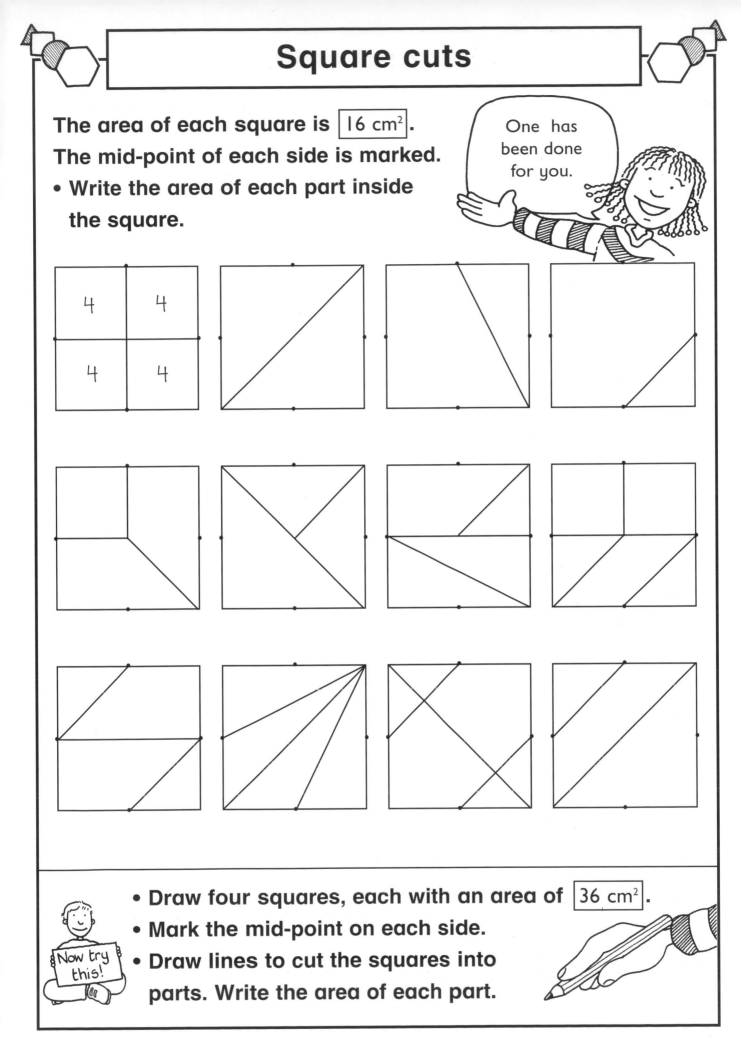

• Draw four squares, each with an area of 36 cm^2.
• Mark the mid-point on each side.
• Draw lines to cut the squares into parts. Write the area of each part.

Teachers' note Show the children how to calculate fractional areas of a square, given the area of the square, and explain to them how the areas of each part can be found by halving squares and rectangles within the square. Explain 'mid-points' if necessary.

Developing Numeracy
Measures, Shape and Space
Year 6
© A & C Black

Polka dot polygons

• **Join the dots to make different polygons.**

There are 19. Try to find them all.

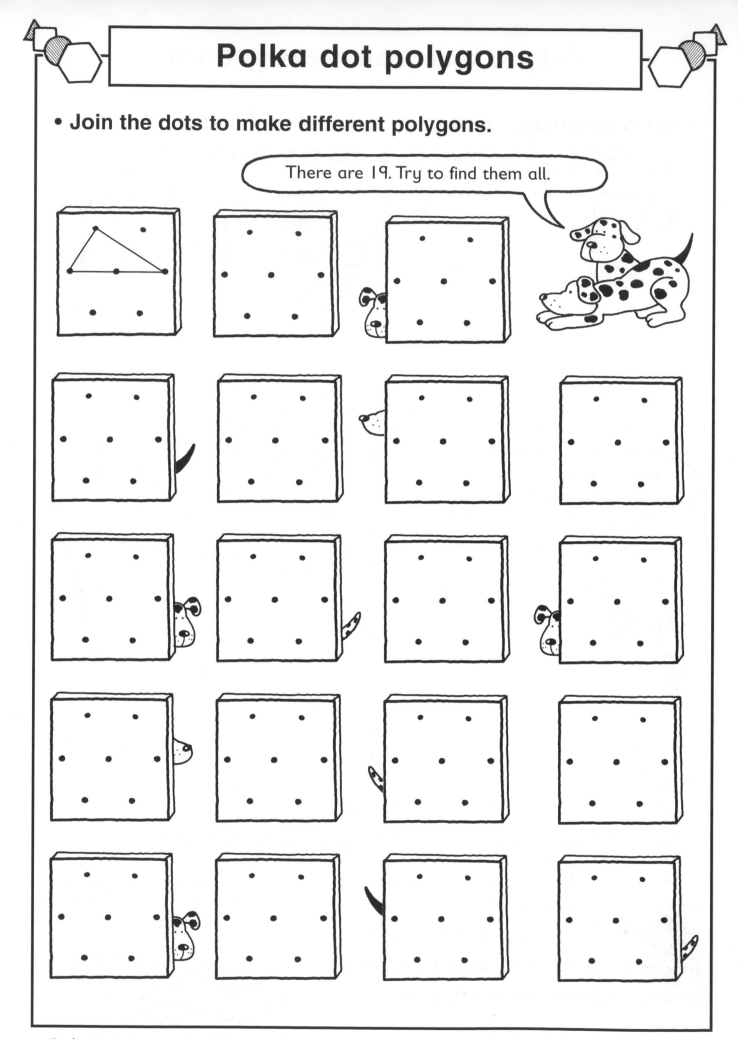

Teachers' note Encourage the children to work systematically by, for example, starting to find all the possible triangles, then quadrilaterals, and so on. Suggest that they draw their polygons in pencil first, or use a geoboard. As an extension, ask the children to write the name of each polygon underneath.

Developing Numeracy
Measures, Shape and Space
Year 6
© A & C Black

Angles of parallelograms

- **Use a protractor to measure the angles of these** parallelograms **. Measure to the nearest degree.**

A **parallelogram** is a quadrilateral with two pairs of parallel sides.

1.

2.

3.

4.

- **What do you notice about the opposite angles of a parallelogram?** _____

Now try this!

- **On squared paper, draw three parallelograms of your own.**
- **Measure the angles.**

Teachers' note As a further extension, ask the children to construct a parallelogram with given angles, for example 60° and 120°.

**Developing Numeracy
Measures, Shape and Space
Year 6
© A & C Black**

Trapezium teaser

- **Cut out the square at the bottom of the page. Cut along the lines.**
- **Join pieces to make different** trapeziums **.**
- **Draw each new shape.**

A **trapezium** is a quadrilateral with only one pair of parallel sides.

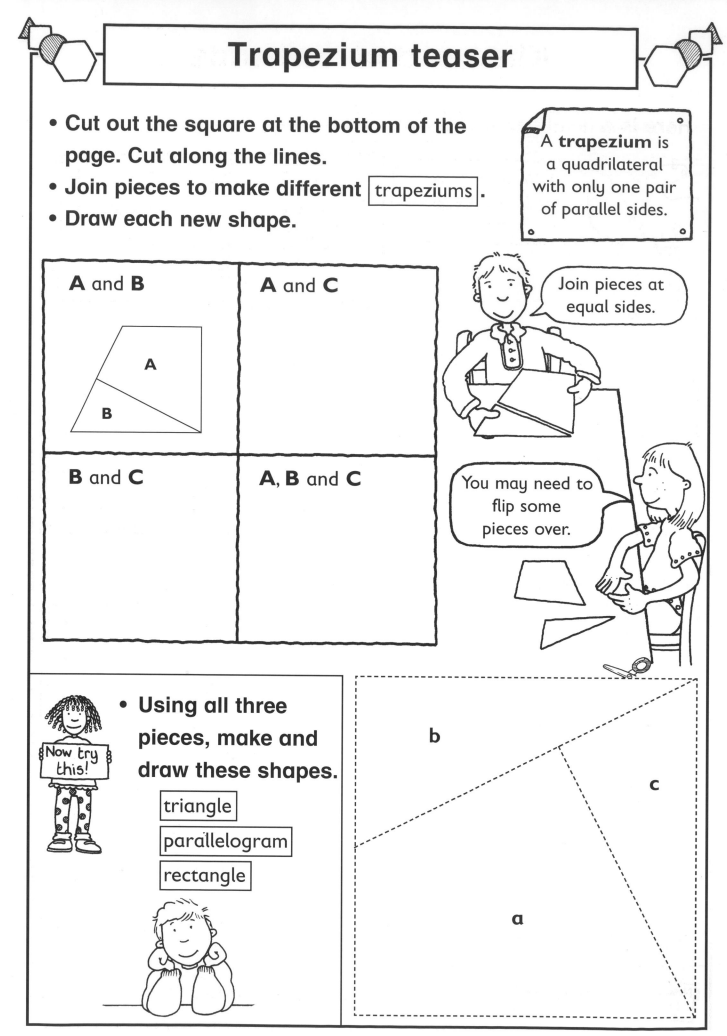

A and **B**	**A** and **C**
B and **C**	**A, B** and **C**

Join pieces at equal sides.

You may need to flip some pieces over.

Now try this!

- **Using all three pieces, make and draw these shapes.**

triangle

parallelogram

rectangle

Teachers' note Encourage the children to cut out the pieces as accurately as possible. Remind them that the pieces must have **equal** sides joined.

Developing Numeracy Measures, Shape and Space Year 6 © A & C Black

Kites and arrowheads

Here is a kite **and an** arrowhead **.**

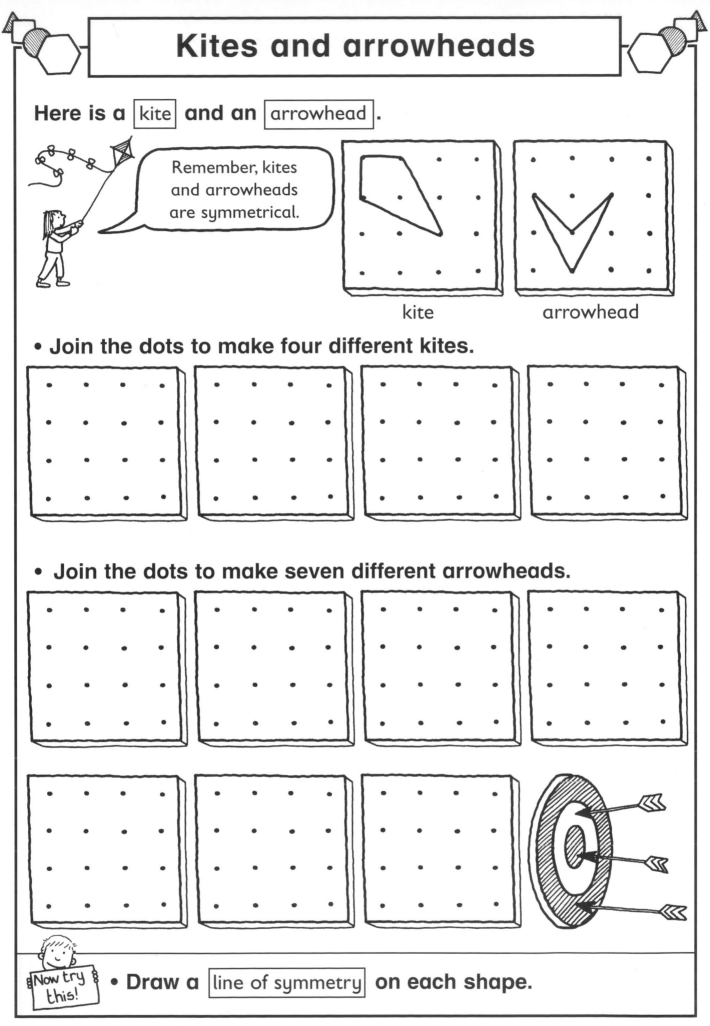

Remember, kites and arrowheads are symmetrical.

kite   arrowhead

• **Join the dots to make four different kites.**

• **Join the dots to make seven different arrowheads.**

• **Draw a** line of symmetry **on each shape.**

Now try this!

Teachers' note Clarify the difference between the two shapes, namely that a kite is convex and an arrowhead is concave, i.e. it contains a reflex angle. Suggest that the children work in pencil first, or use a geoboard.

Developing Numeracy
Measures, Shape and Space
Year 6
© A & C Black

33

Quadrilateral challenge

- **Join the dots to make** quadrilaterals .

Can you find 16 different ones?

Some of the quadrilaterals have special names:

parallelogram · rhombus · trapezium · kite · arrowhead

- **Write the correct names underneath each shape.**

Teachers' note Encourage the children to work in pencil first. They could use a geoboard to create different quadrilaterals, and to check whether or not two are identical. As an extension activity, ask the children to find the area of each quadrilateral.

Developing Numeracy
Measures, Shape and Space
Year 6
© A & C Black

34

Quadrilateral check-list

- **Complete the check-list. Put:**
  - ✓ if the statement is **always** true
  - ? if the statement is **sometimes** true
  - ✗ if the statement is **never** true

	rectangle	square	parallelogram	rhombus	trapezium	kite	arrowhead
4 sides	✓						
all sides equal							
2 pairs of equal sides							
4 angles							
all angles equal							
2 pairs of equal angles							
at least 1 pair of parallel sides							
at least 1 pair of perpendicular sides							
at least 1 right angle							
line symmetry							

- **Create a check-list for triangles. Use these triangles.**

| equilateral | isosceles | right-angled | scalene |

Teachers' note The children could work in pairs. They will find it helpful to have at least one of each shape to look at. Encourage them to discuss the properties of the shapes in detail before marking their answers on the chart. The activity provides an opportunity to discuss the fact that a rectangle is also a parallelogram and a square is also a rhombus.

Developing Numeracy
Measures, Shape and Space
Year 6
© A & C Black

Quadrilateral quandary

• **Play this game with a partner.**

You need five counters each (in two colours) and two dice.

☆ Take turns to throw two dice. Add the numbers.

☆ Place a counter on the shape with that number. If the shape already has a counter on it, miss a go.

☆ Say the name of the shape. If your partner agrees, leave your counter in place. Otherwise, remove it.

☆ The winner is the first to use all five counters.

rectangle | arrowhead

square | rhombus | kite

parallelogram | trapezium

2 3 4

5 6 7

8 9

10 11 12

Teachers' note Each pair of children needs one copy of the sheet. If necessary, revise the definition of each shape listed on the wall. As an extension activity, ask the children to list the properties of each shape in terms of angles, parallel sides, and so on.

Developing Numeracy
Measures, Shape and Space
Year 6
© A & C Black

Drawing diagonals

- **Draw both diagonals on each quadrilateral.**

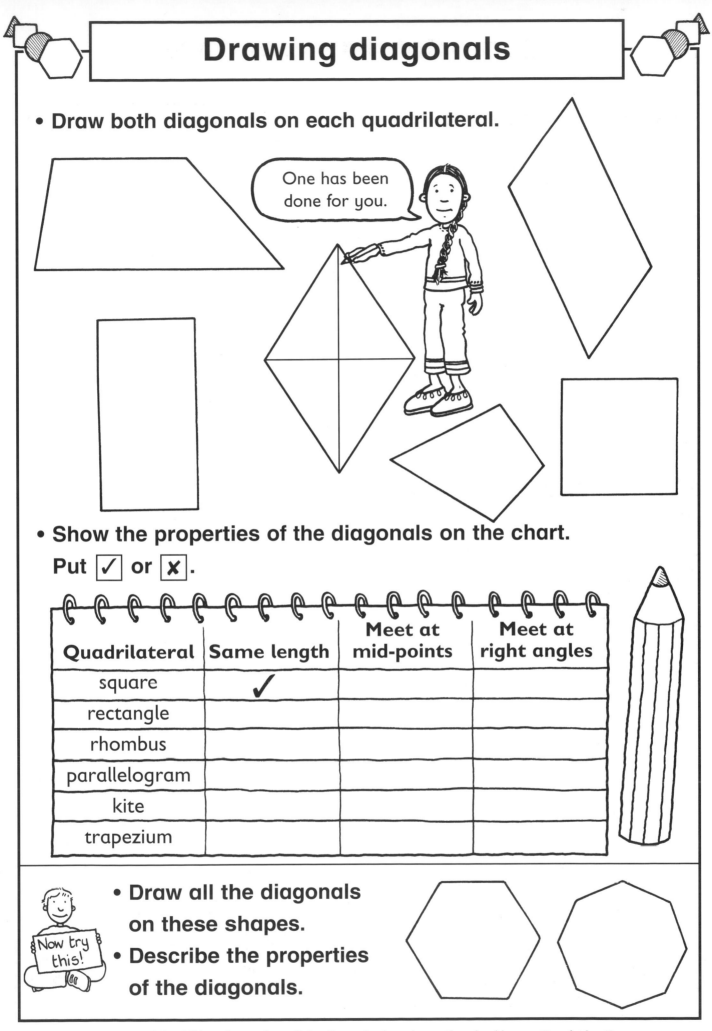

One has been done for you.

- **Show the properties of the diagonals on the chart.**

Put ☑ or ☒.

Quadrilateral	Same length	Meet at mid-points	Meet at right angles
square	✓		
rectangle			
rhombus			
parallelogram			
kite			
trapezium			

Now try this!

- **Draw all the diagonals on these shapes.**
- **Describe the properties of the diagonals.**

Teachers' note Remind the children that to draw all the diagonals of a polygon, they should draw a straight line from each vertex to every other vertex. They can use a set-square to test for right angles. Discuss 'mid-points' if necessary.

Developing Numeracy
Measures, Shape and Space
Year 6
© A & C Black

Fly the flag

- Draw two straight lines on each flag to make four shapes.
- Find different shapes. Write their names.

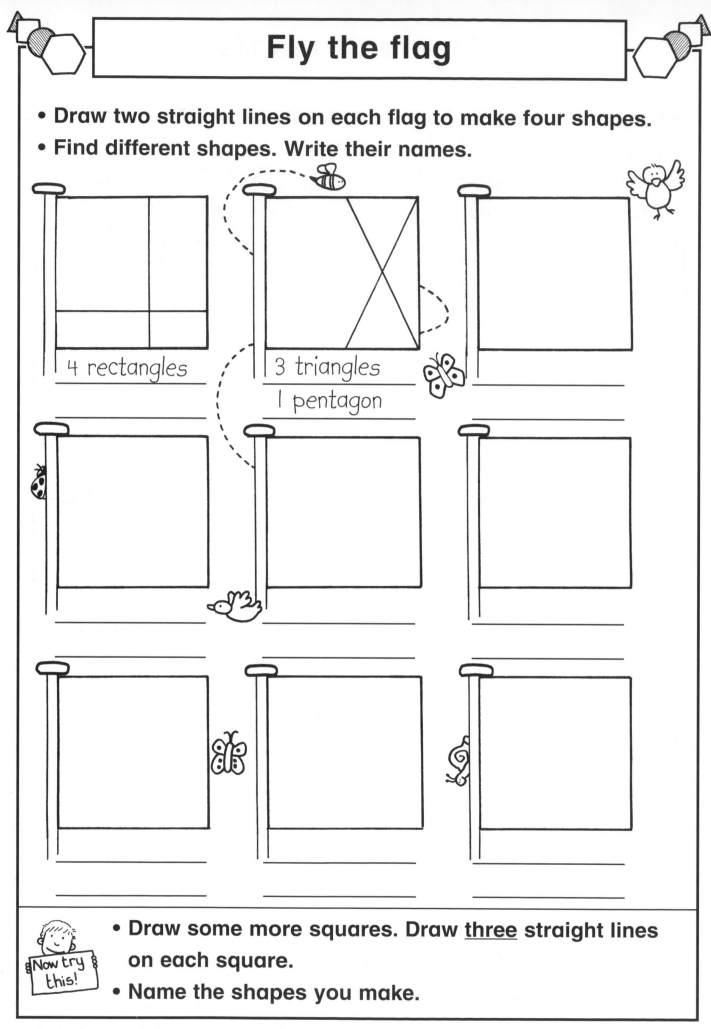

4 rectangles

3 triangles
1 pentagon

- Draw some more squares. Draw <u>three</u> straight lines on each square.
- Name the shapes you make.

Teachers' note As a further extension, you could ask the children to draw lines on a different shape, for example, a regular hexagon.

**Developing Numeracy
Measures, Shape and Space
Year 6
© A & C Black**

Five-piece puzzle

- **Cut out the square at the bottom. Cut along the lines.**
- **Join** [A] , [B] **and** [C] **along equal sides to make new shapes.**
- **Draw the new shapes.**

Flip the shape if you need to!

rectangle	right-angled triangle	parallelogram	trapezium	pentagon
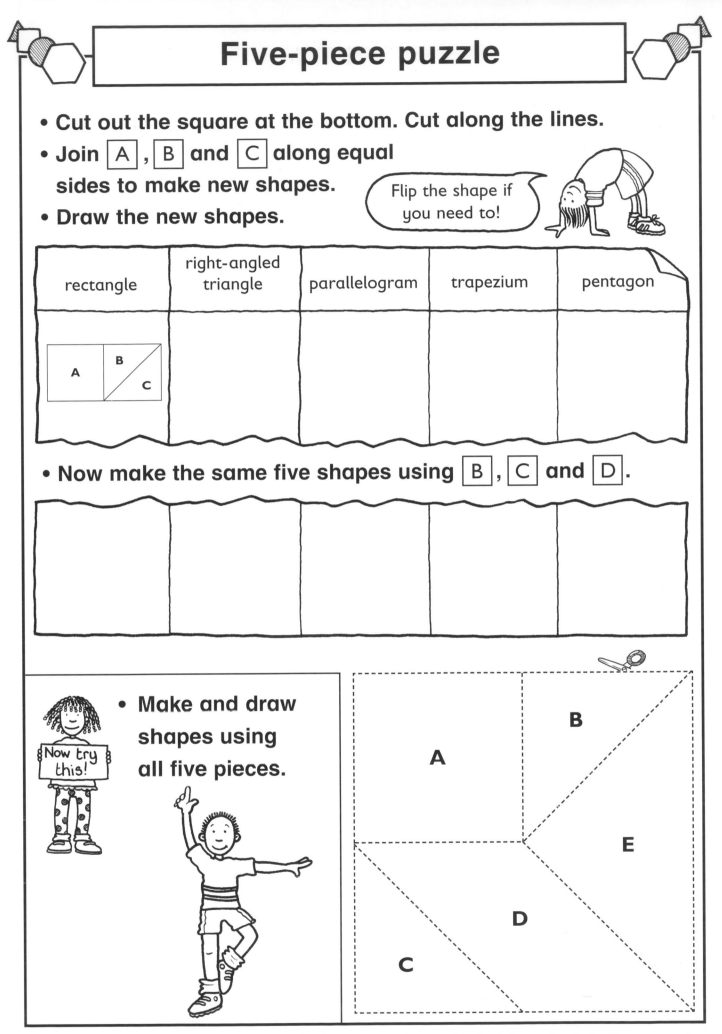				

- **Now make the same five shapes using** [B] , [C] **and** [D] .

- **Make and draw shapes using all five pieces.**

Now try this!

Teachers' note Encourage the children to cut out the pieces as accurately as possible. Remind them that the pieces must have **equal** sides joined.

Developing Numeracy
Measures, Shape and Space
Year 6
© A & C Black

39

Hexagon puzzle

- **Cut out the hexagon at the bottom. Cut along the lines.**
- **Cut out the cards. Put them in a pile.**
- **Take turns to pick a card and make the shape.**

Work with a partner.

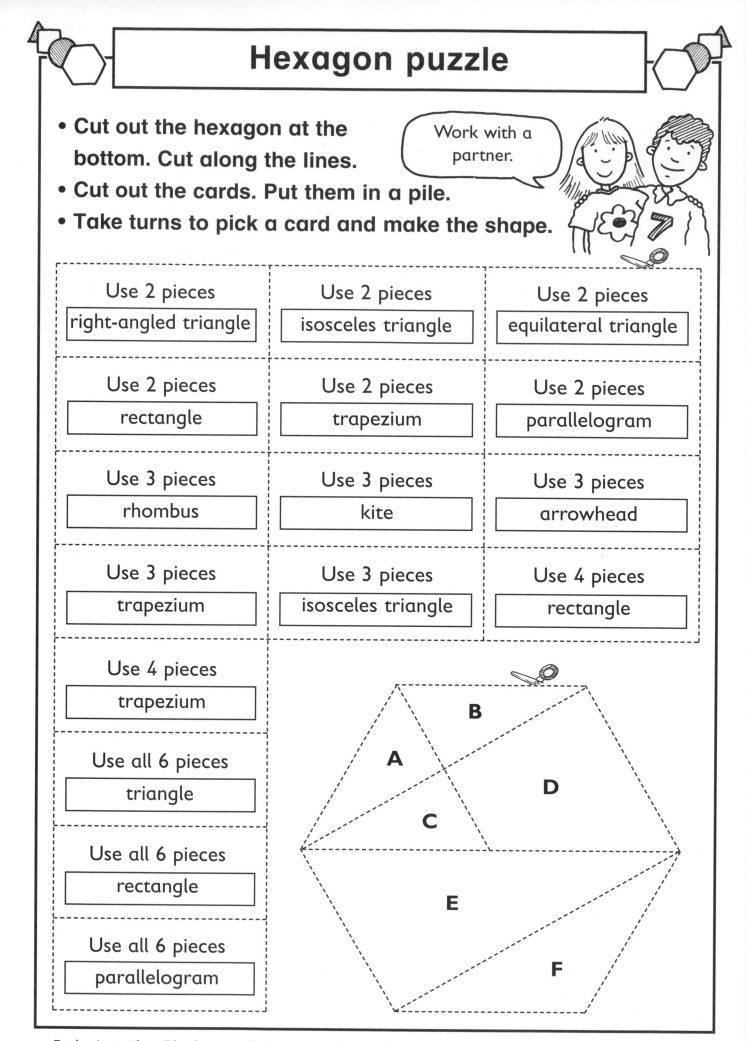

Use 2 pieces	Use 2 pieces	Use 2 pieces
right-angled triangle	isosceles triangle	equilateral triangle

Use 2 pieces	Use 2 pieces	Use 2 pieces
rectangle	trapezium	parallelogram

Use 3 pieces	Use 3 pieces	Use 3 pieces
rhombus	kite	arrowhead

Use 3 pieces	Use 3 pieces	Use 4 pieces
trapezium	isosceles triangle	rectangle

Use 4 pieces
trapezium

Use all 6 pieces
triangle

Use all 6 pieces
rectangle

Use all 6 pieces
parallelogram

Teachers' note If possible, photocopy the sheet onto card. Before beginning the activity, discuss the properties of the six shapes which make up the hexagon. The children can record the shapes they create by drawing round the pieces on a large sheet of paper. They could show the joins with dotted lines. Remind the children that they can flip shapes.

Developing Numeracy
Measures, Shape and Space
Year 6
© A & C Black

Shady shapes

- **Shade part of each pentagon to create these shapes.**

Each shape must be different.

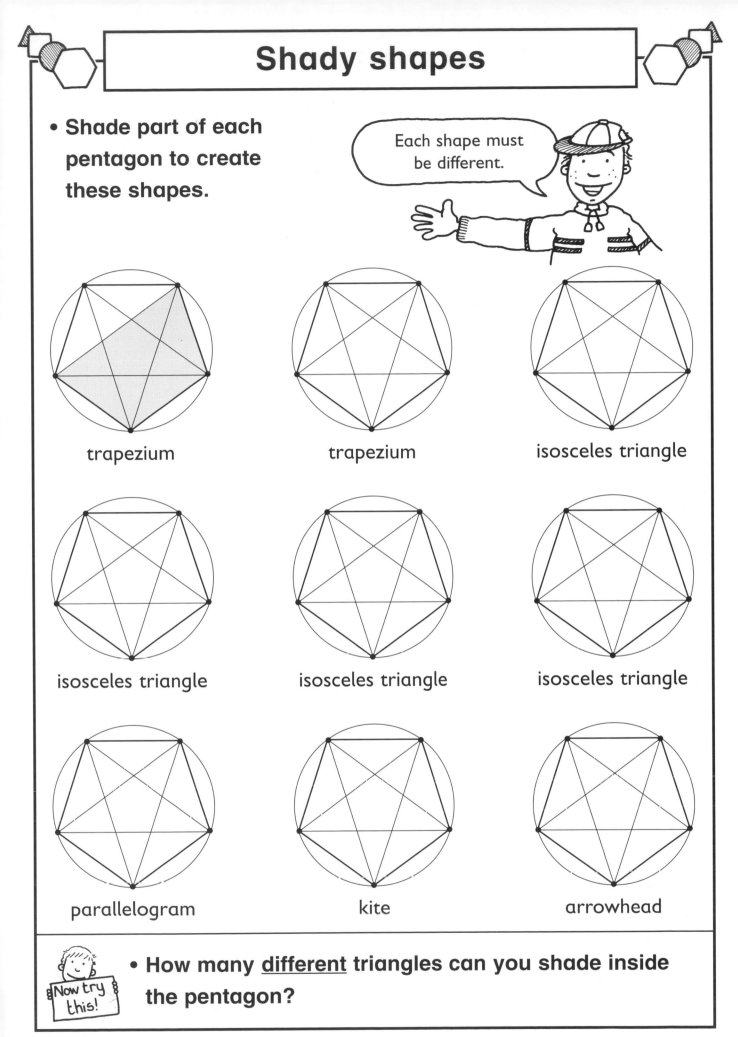

trapezium

trapezium

isosceles triangle

isosceles triangle

isosceles triangle

isosceles triangle

parallelogram

kite

arrowhead

- **How many different triangles can you shade inside the pentagon?**

Now try this!

Teachers' note Emphasise that each shape must be different, i.e. not congruent. For the extension activity children may need more copies of the sheet.

Developing Numeracy
Measures, Shape and Space
Year 6
© A & C Black

The shape game

- **Play this game with a partner.**

☆ Cut out the cards. Spread them face down.

☆ Take turns to reveal a card. Describe the angles of the shape.

Example: 2 equal acute, 2 equal obtuse

☆ If your partner agrees, keep the card. Otherwise, replace it.

☆ The winner is the player with the most cards.

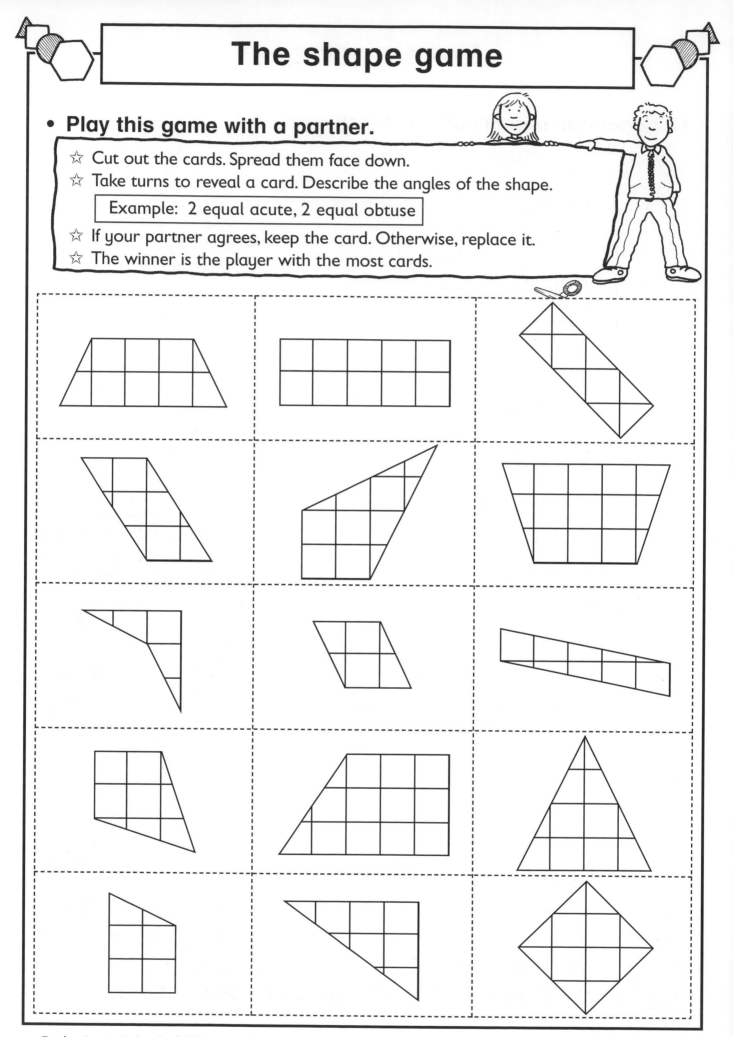

Teachers' note Each pair of children needs one copy of the sheet. The children should agree on the correct answers together. Any they are not sure of should be put to one side and discussed in the plenary session. The cards can also be used for describing other properties, for example, shapes which have parallel sides, shapes which have equal sides; or as a shape naming game.

Developing Numeracy
Measures, Shape and Space
Year 6
© A & C Black

Net the shape

Here are the nets of some solid shapes.

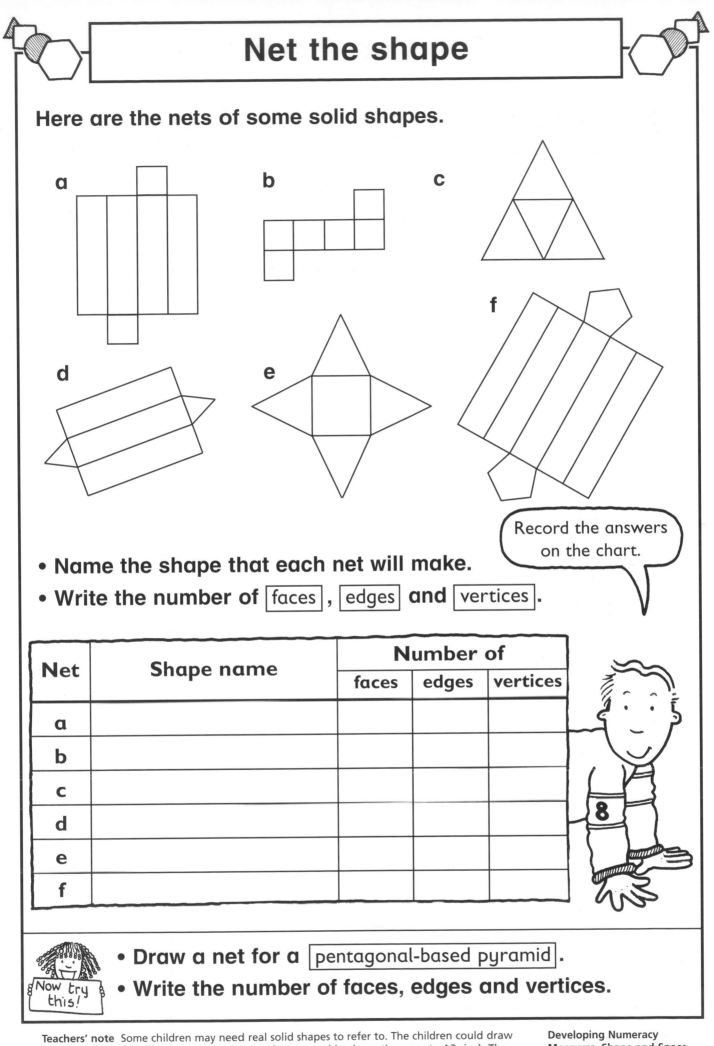

a

b

c

f

d

e

Record the answers on the chart.

- **Name the shape that each net will make.**
- **Write the number of** faces **,** edges **and** vertices **.**

Net	Shape name	Number of		
		faces	edges	vertices
a				
b				
c				
d				
e				
f				

Now try this!

- **Draw a net for a** pentagonal-based pyramid **.**
- **Write the number of faces, edges and vertices.**

Teachers' note Some children may need real solid shapes to refer to. The children could draw larger versions of each net on squared paper (or you could enlarge the page to A3 size). The children can then draw tabs on alternate sides around each net, cut out the nets and construct the 3-D shapes. For the extension activity, provide pentagon templates.

**Developing Numeracy
Measures, Shape and Space
Year 6
© A & C Black**

Spot on!

On a dice, the spots on opposite faces always total $\boxed{7}$.

- Look at these nets of dice. Some of the spots are missing.
- Draw the missing spots.

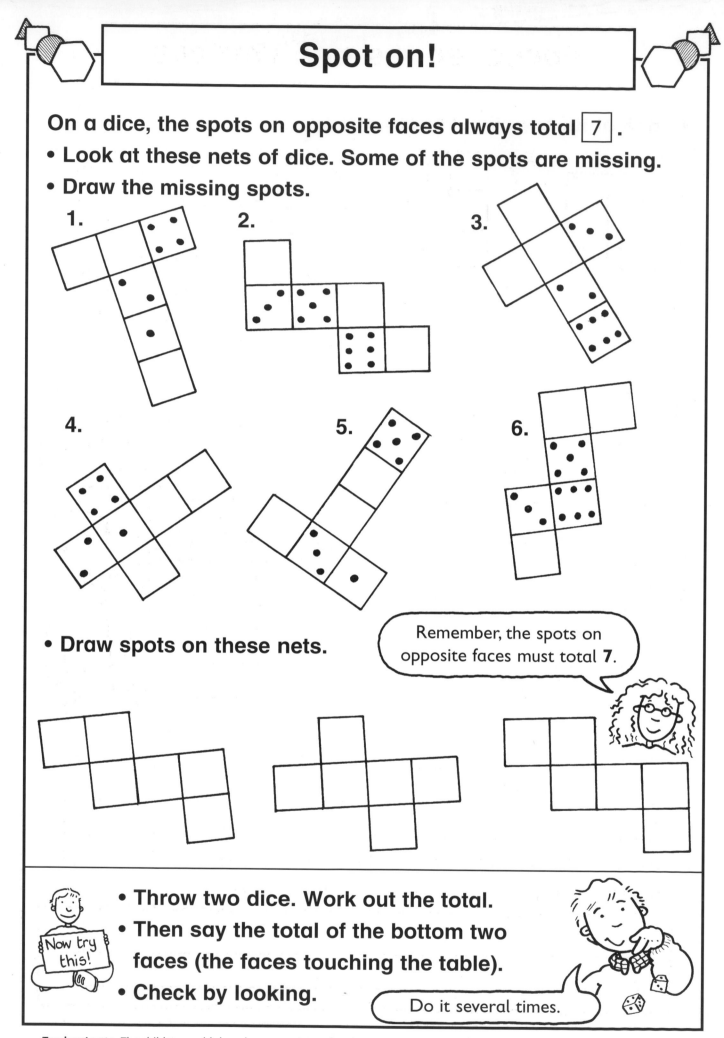

1.

2.

3.

4.

5.

6.

- Draw spots on these nets.

Remember, the spots on opposite faces must total **7**.

Teachers' note The children could draw larger versions of each net on squared paper (or you could enlarge the page to A3 size). The children can then draw tabs on alternate sides around each net, cut out the nets and fold to check that opposite faces total 7.

Developing Numeracy
Measures, Shape and Space
Year 6
© A & C Black

Faces, edges and vertices

- **Look at these solid shapes.**

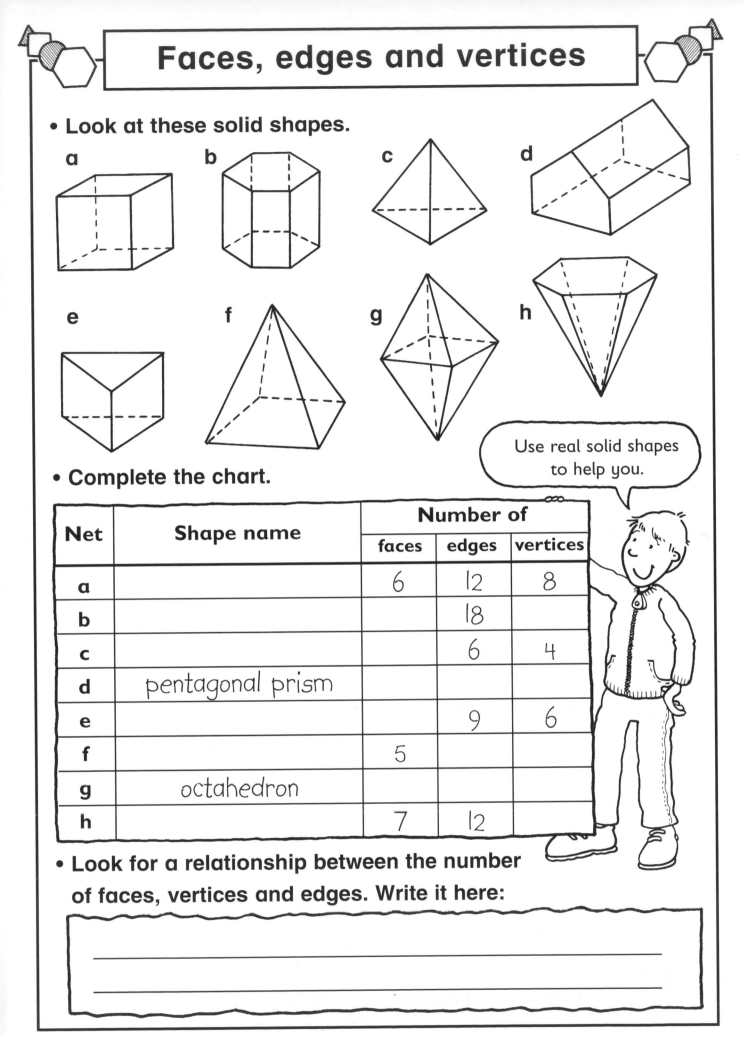

a   b   c   d

e   f   g   h

Use real solid shapes to help you.

- **Complete the chart.**

Net	Shape name	Number of		
		faces	edges	vertices
a		6	12	8
b			18	
c			6	4
d	pentagonal prism			
e			9	6
f		5		
g	octahedron			
h		7	12	

- **Look for a relationship between the number of faces, vertices and edges. Write it here:**

Teachers' note If possible, provide real solid shapes for reference. If children struggle to find a relationship, suggest that they total the number of faces and vertices, and compare this with the number of edges. As an extension, ask the children to investigate the number of parallel faces each shape has, and which of the shapes have perpendicular faces.

Developing Numeracy
Measures, Shape and Space
Year 6
© A & C Black

Tetrahedron and octahedron

- **Follow the instructions.**

☆ On each net, fill in the number of edges, faces and vertices of the made-up shape.

☆ Cut out the nets. Score along the lines with a pencil and ruler.

☆ Fold and glue the nets to make the shapes.

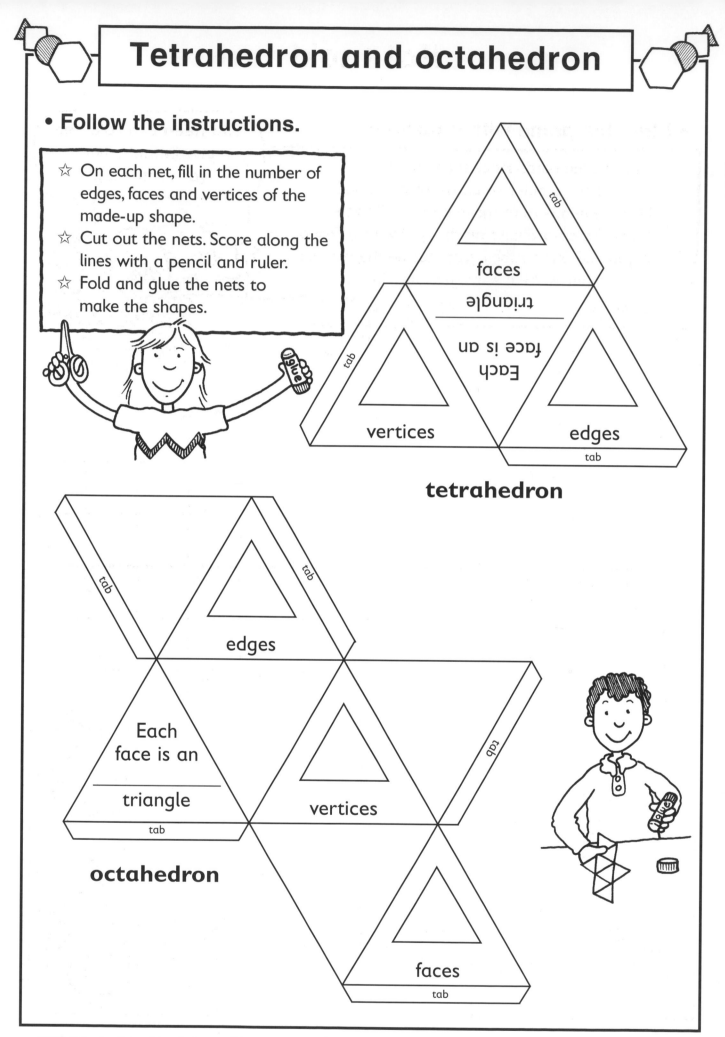

tetrahedron

octahedron

Teachers' note If possible, photocopy this page onto card or A3 paper to make the nets easier to construct.

**Developing Numeracy
Measures, Shape and Space
Year 6
© A & C Black**

Hat trick!

• **Play this game with a partner.**

☆ Cut out the cards. Place them in a hat.
☆ Take turns to pull two cards from the hat.
  Use the numbers to make a **co-ordinate**.
☆ Mark the co-ordinate on the grid with a cross.
  If you pull out a rabbit, **you** decide the number.
☆ The winner is the first to get three in a line.

You need two coloured pencils and a hat (or container).

Game 1

Game 2

Teachers' note Each pair of children needs one copy of the sheet. They may need reminding of how to plot co-ordinates. The game can be played in two ways: either the children pull out the cards one at a time – the first card gives the horizontal co-ordinate and the second the vertical; or they pull out both cards together and choose which card represents which co-ordinate.

**Developing Numeracy
Measures, Shape and Space
Year 6
© A & C Black**

Alien co-ordinates

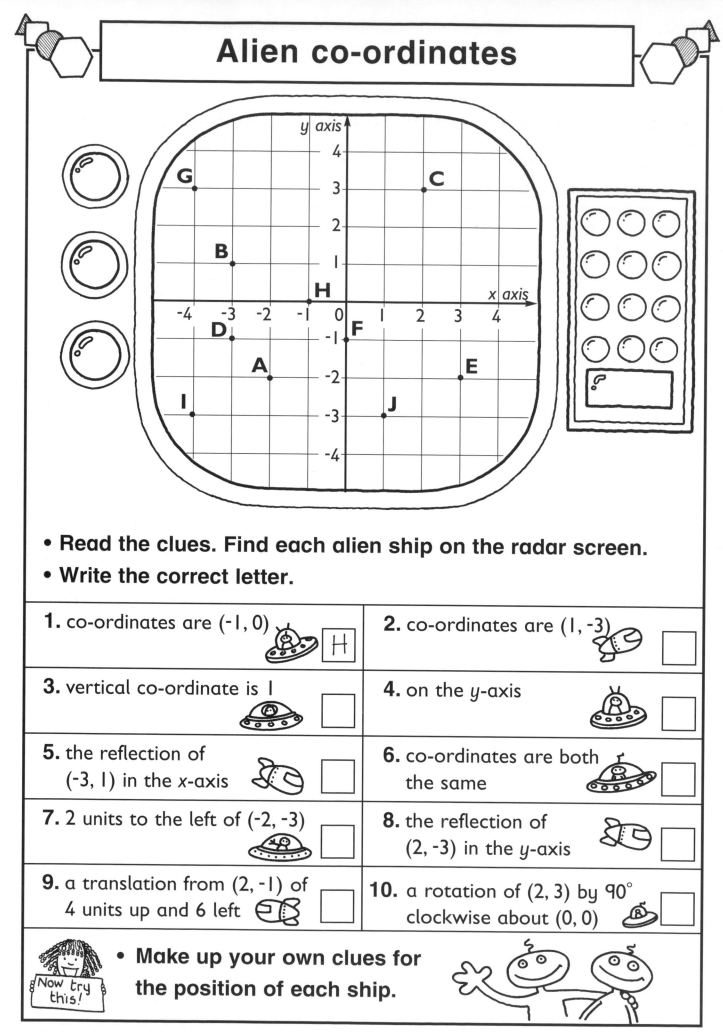

- **Read the clues. Find each alien ship on the radar screen.**
- **Write the correct letter.**

1. co-ordinates are (-1, 0) $\boxed{\text{H}}$

2. co-ordinates are (1, -3) $\square$

3. vertical co-ordinate is 1 $\square$

4. on the y-axis $\square$

5. the reflection of (-3, 1) in the x-axis $\square$

6. co-ordinates are both the same $\square$

7. 2 units to the left of (-2, -3) $\square$

8. the reflection of (2, -3) in the y-axis $\square$

9. a translation from (2, -1) of 4 units up and 6 left $\square$

10. a rotation of (2, 3) by 90° clockwise about (0, 0) $\square$

- **Make up your own clues for the position of each ship.**

Now try this!

Teachers' note Ensure that the children understand reflections, rotations and translations. If necessary, revise 'x-axis' and 'y-axis'. In the extension activity, the children could give their clues to a partner to solve.

Developing Numeracy
Measures, Shape and Space
Year 6
© A & C Black

Co-ordinate crazy!

- **Join the points in order. Write the name of the quadrilateral.**
- **On the final grid, draw your own quadrilateral. Write the co-ordinates and its name.**

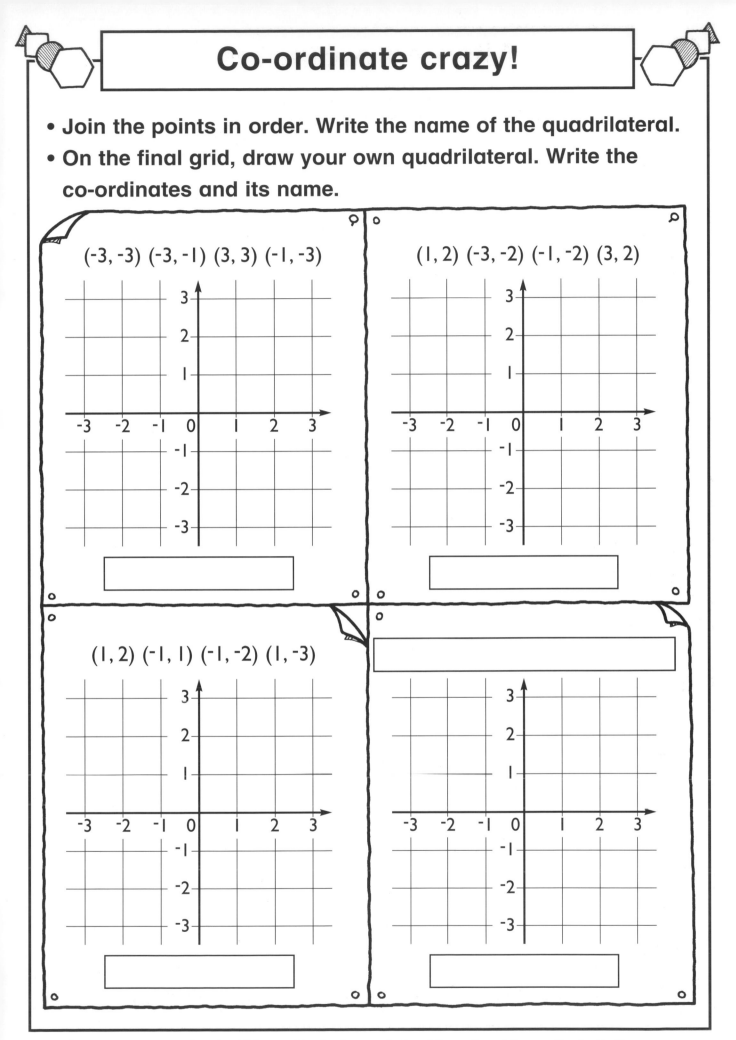

(-3, -3) (-3, -1) (3, 3) (-1, -3)

(1, 2) (-3, -2) (-1, -2) (3, 2)

(1, 2) (-1, 1) (-1, -2) (1, -3)

Teachers' note As an extension, the children could write the co-ordinates of the vertices of three more quadrilaterals, making each shape symmetrical. They could then rotate each shape about the origin, write the new co-ordinates of the position of the vertices, and compare them with the co-ordinates of the original shape.

Developing Numeracy
Measures, Shape and Space
Year 6
© A & C Black

Symmetry of quadrilaterals

- **Draw all the** lines of symmetry **on the quadrilaterals.**
- **Complete the chart.**

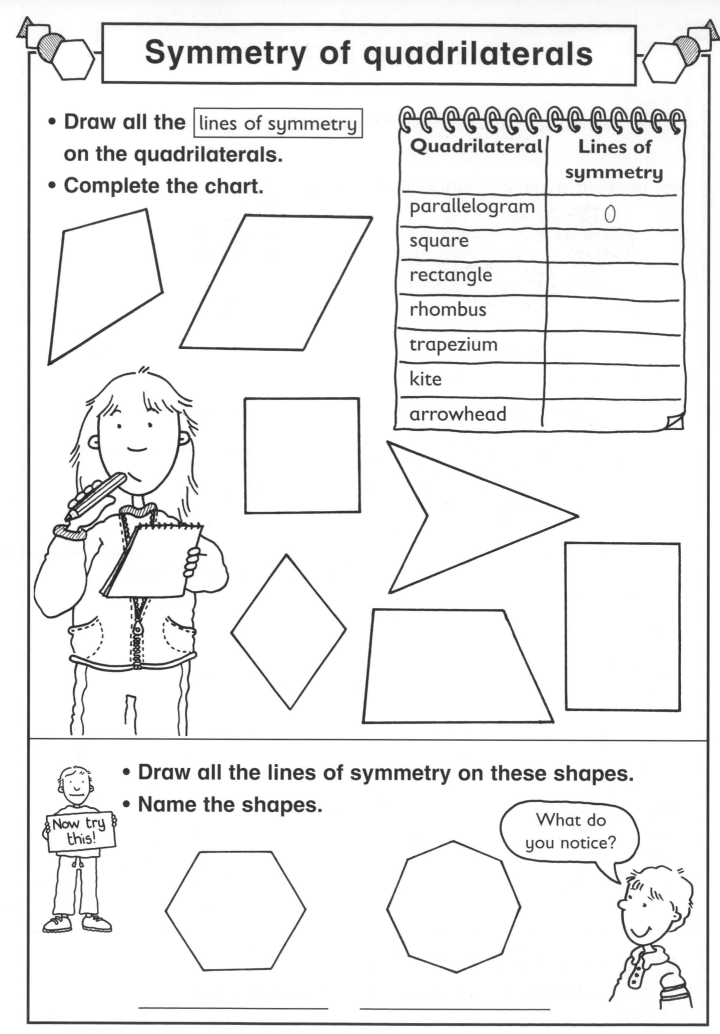

Quadrilateral	Lines of symmetry
parallelogram	0
square	
rectangle	
rhombus	
trapezium	
kite	
arrowhead	

- **Draw all the lines of symmetry on these shapes.**
- **Name the shapes.**

Now try this!

What do you notice?

_____   _____

Teachers' note Explain to the children that some of the shapes do not have any lines of symmetry.

Developing Numeracy
Measures, Shape and Space
Year 6
© A & C Black

Rotation patterns

Chloe is using $\boxed{\text{rotation}}$ **to make patterns.**

She draws a shape and marks a vertex.

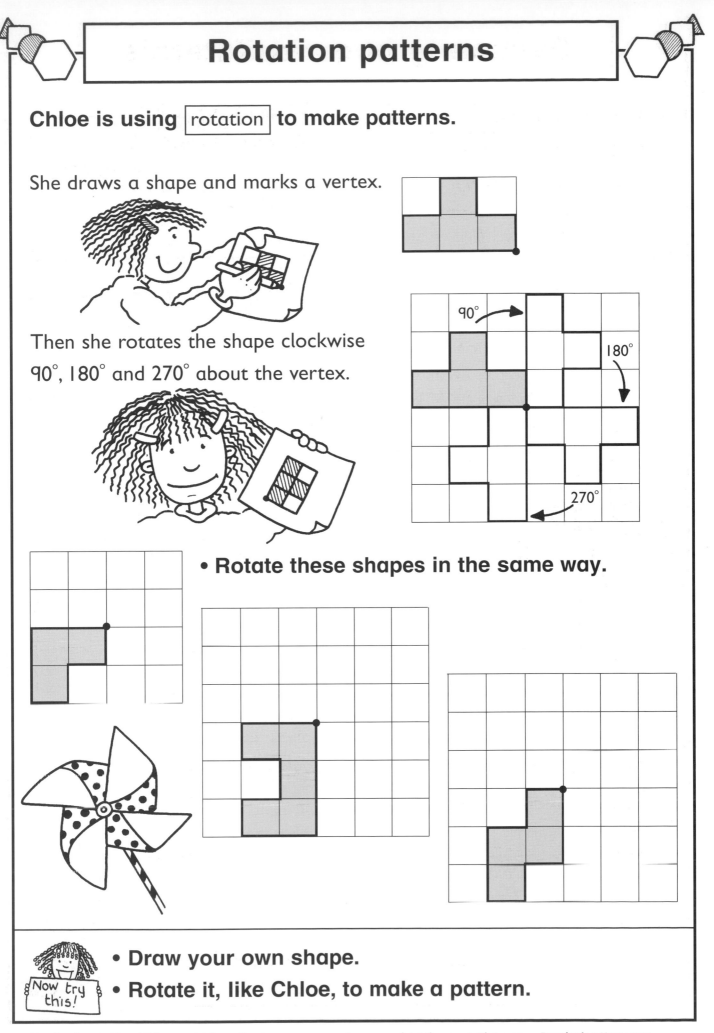

Then she rotates the shape clockwise 90°, 180° and 270° about the vertex.

• **Rotate these shapes in the same way.**

• **Draw your own shape.**
• **Rotate it, like Chloe, to make a pattern.**

Teachers' note The children could draw the shapes on squared paper and cut them out. They can then rotate them about the suggested points, before drawing the new position on the grid.

**Developing Numeracy
Measures, Shape and Space
Year 6**
© A & C Black

Rotating shapes

- **Draw the shape after each rotation.**
- **Write the new co-ordinates.**

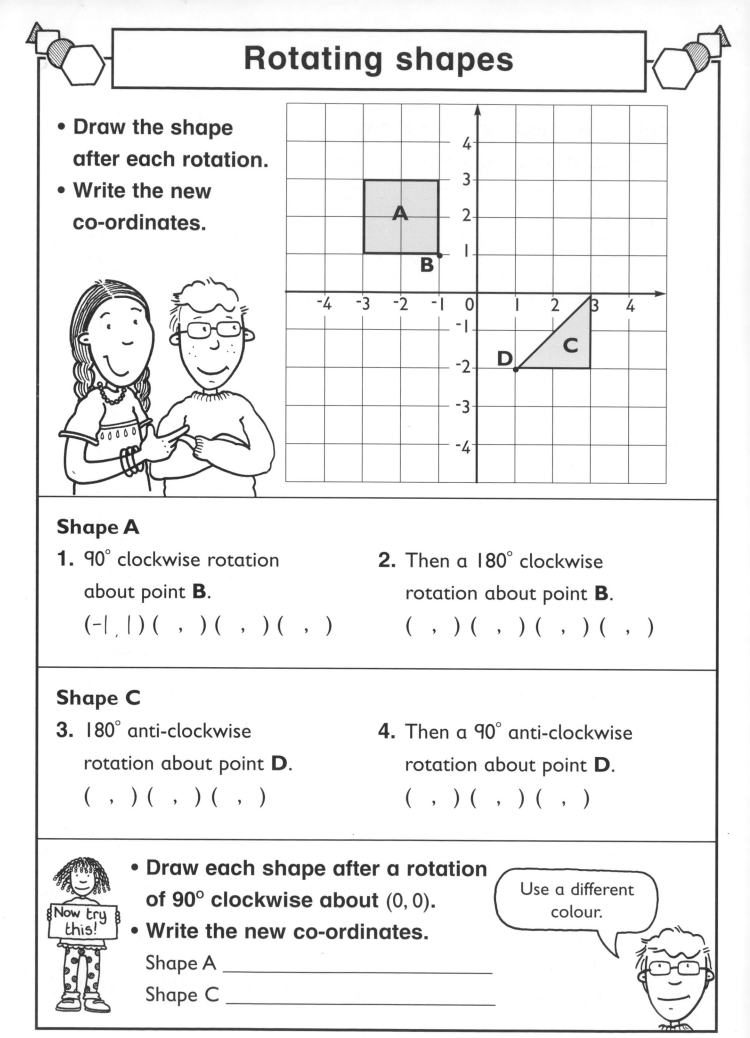

Shape A

1. 90° clockwise rotation about point **B**.

(-1, 1) (,) (,) (,)

2. Then a 180° clockwise rotation about point **B**.

(,) (,) (,) (,)

Shape C

3. 180° anti-clockwise rotation about point **D**.

(,) (,) (,)

4. Then a 90° anti-clockwise rotation about point **D**.

(,) (,) (,)

Now try this!

- **Draw each shape after a rotation of 90° clockwise about** (0, 0).
- **Write the new co-ordinates.**

Shape A _____

Shape C _____

Use a different colour.

Teachers' note The children can draw a large co-ordinate grid, draw and cut out card or paper models of each shape, then rotate them on the grid to check their positions after rotation.

Developing Numeracy Measures, Shape and Space Year 6 © A & C Black

Moth moves

- **Look at the moth on the window.**
- **Draw the moth after each** translation .

Start from the original position each time.

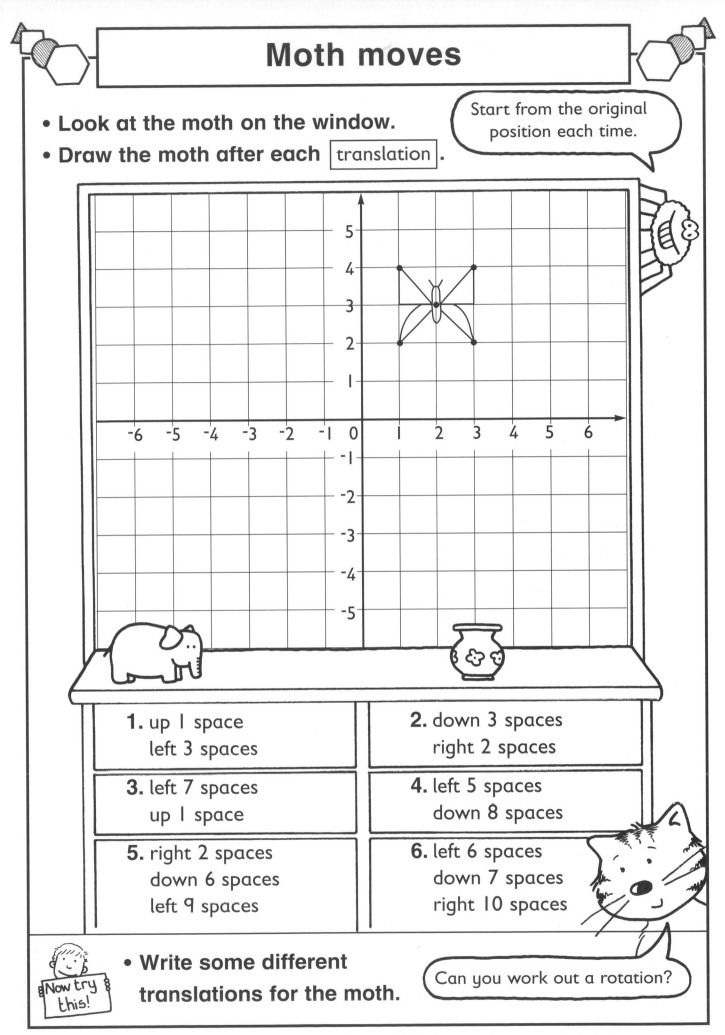

1. up 1 space
    left 3 spaces

2. down 3 spaces
    right 2 spaces

3. left 7 spaces
    up 1 space

4. left 5 spaces
    down 8 spaces

5. right 2 spaces
    down 6 spaces
    left 9 spaces

6. left 6 spaces
    down 7 spaces
    right 10 spaces

Now try this!

- **Write some different translations for the moth.**

Can you work out a rotation?

Teachers' note Suggest that the children number each translation to distinguish them. The children could also draw and describe some translations using compass directions.

**Developing Numeracy
Measures, Shape and Space
Year 6
© A & C Black**

Seagulls

• **Follow the instructions.**

☆ Estimate the angle. Then measure it.

☆ Find the difference between the two. This is your score.

☆ Work out your **total** score. Compare it with a partner.

Angle	Estimate	Measure	Score
a			
b			
c			
d			
e			
f			
Total			

A low score is a good score!

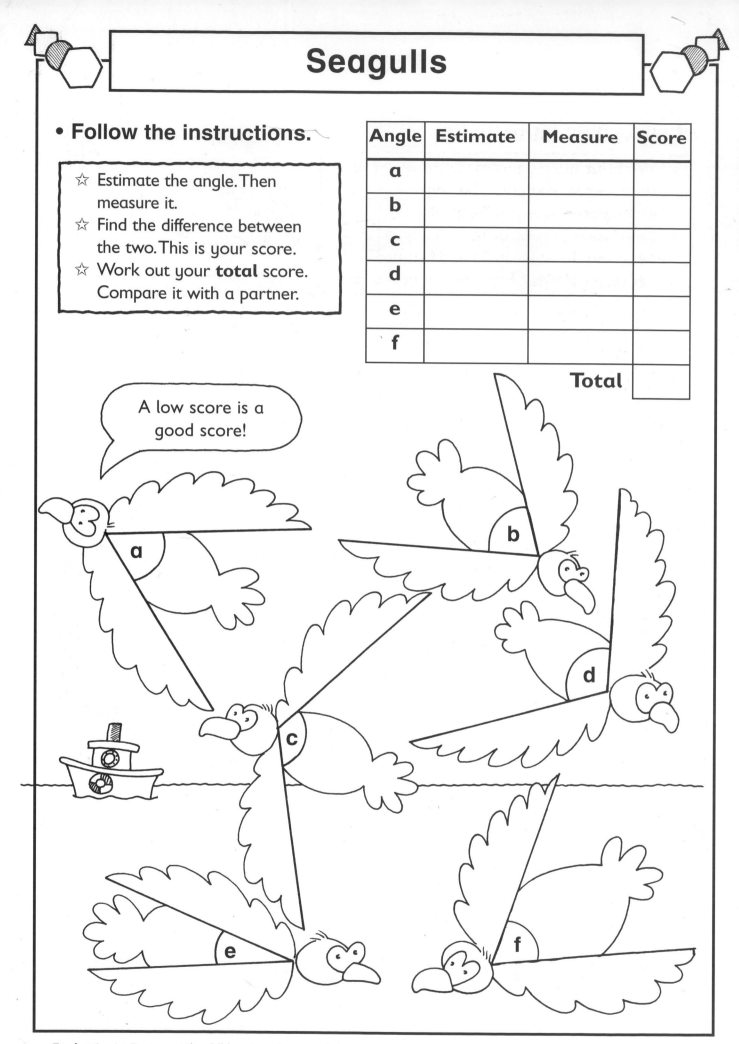

Teachers' note Encourage the children to estimate and then measure one angle at a time to help them to gradually improve their estimates. Suggest that they think about each angle in relation to a right angle when making their estimates. Stress the need to accurately line up the protractor along one 'arm' of the angle when measuring.

Developing Numeracy
Measures, Shape and Space
Year 6
© A & C Black

Pass!

- **Follow the instructions.**

  ☆ **Without** using a protractor, draw a line to show the next pass. The angle of the pass is given in the circle.

  ☆ Now use a protractor to check your angle.

  ☆ Write on the score ball the difference between your estimate and the actual angle.

Example:

25°

A low score is a good score!

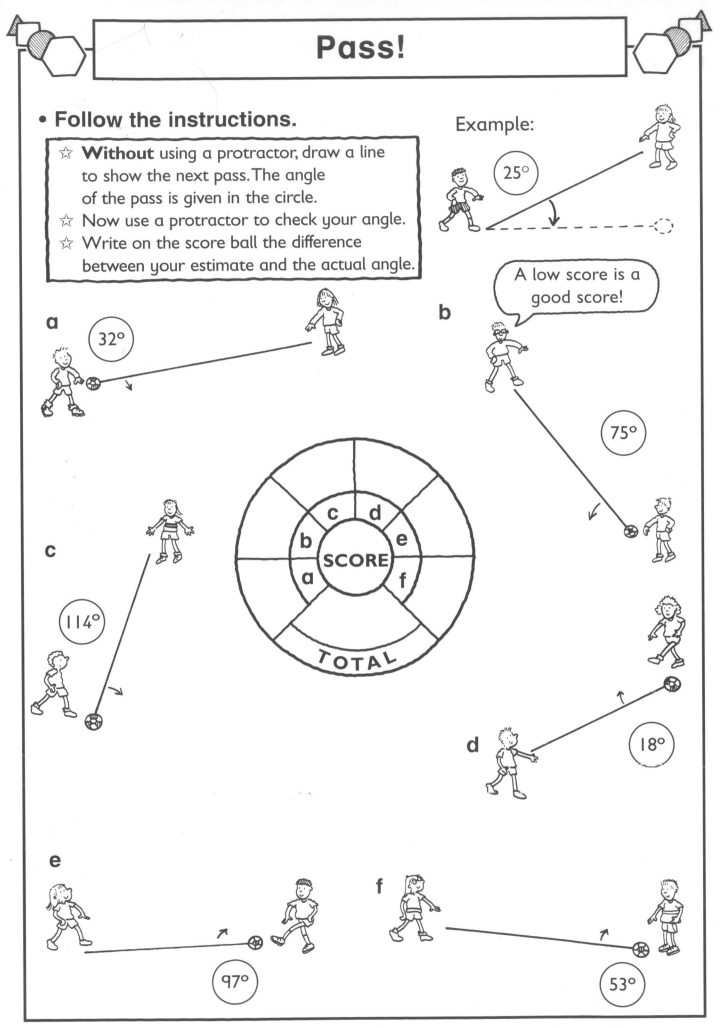

a 32°

b 75°

c 114°

d 18°

e 97°

f 53°

c d
b e
SCORE
a f

TOTAL

Teachers' note The children could work in pairs, checking each other's measurements before recording the scores. They could compete with each other for the lowest score.

Developing Numeracy
Measures, Shape and Space
Year 6
© A & C Black

Angles of a triangle

- **Measure the angles of each triangle.**
- **Find the total of the three angles for each triangle.**

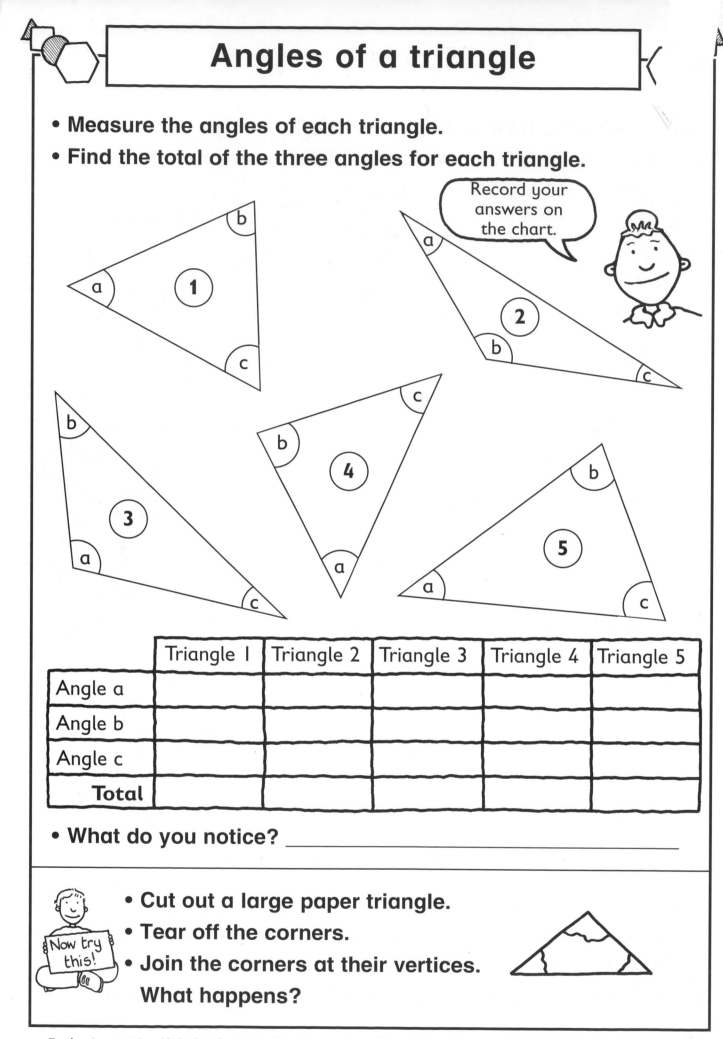

Record your answers on the chart.

	Triangle 1	Triangle 2	Triangle 3	Triangle 4	Triangle 5
Angle a					
Angle b					
Angle c					
Total					

- **What do you notice?** _____

Now try this!

- **Cut out a large paper triangle.**
- **Tear off the corners.**
- **Join the corners at their vertices. What happens?**

Teachers' note It is unlikely that the totals of the three angles will be exactly 180°, because of the difficulty of measuring accurately. As a further extension, the children could draw their own triangles and total the angles to test the theory.

**Developing Numeracy
Measures, Shape and Space
Year 6
© A & C Black**

Mouse party

The mice have nibbled the sandwiches!

• **Write the size of the missing angle on each sandwich.**

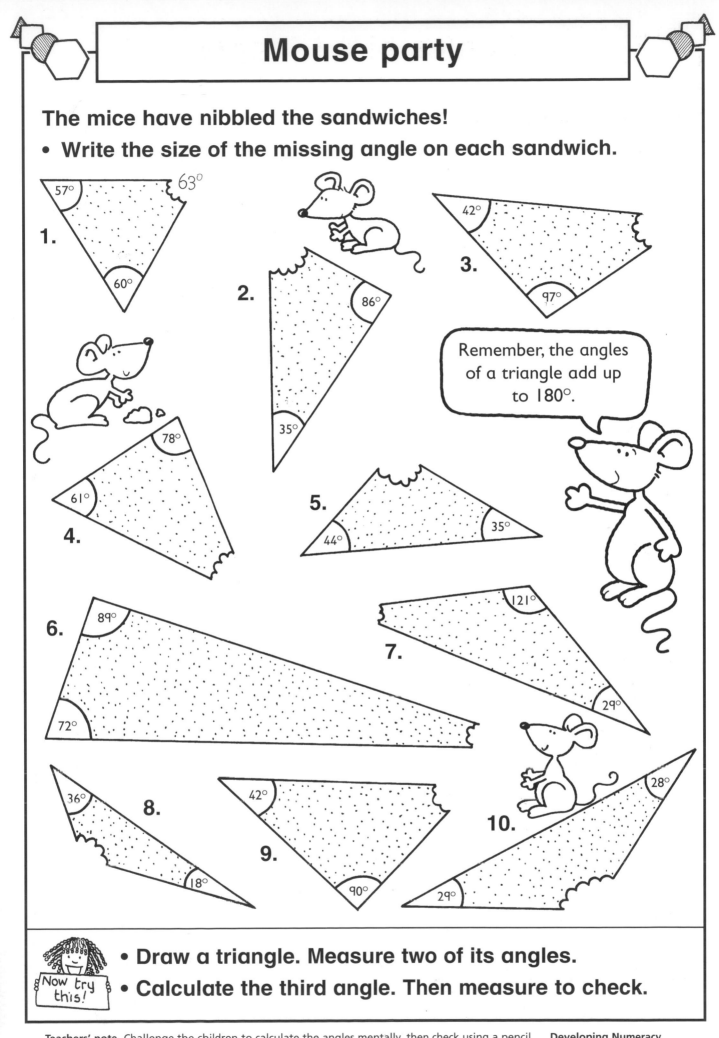

1. 57° 63° 60°

2. 86° 35°

3. 42° 97°

Remember, the angles of a triangle add up to 180°.

4. 78° 61°

5. 44° 35°

6. 89° 72°

7. 121° 29°

8. 36° 18°

9. 42° 90°

10. 28° 29°

Now try this!
• **Draw a triangle. Measure two of its angles.**
• **Calculate the third angle. Then measure to check.**

Teachers' note Challenge the children to calculate the angles mentally, then check using a pencil and paper.

Developing Numeracy
Measures, Shape and Space
Year 6
© A & C Black

Speedy angles

• **Play this game with a partner.**

☆ Start the timer.
☆ Calculate each shaded angle.
Record it on the triangle.
☆ Stop after five minutes. Compare
your answers with your partner's.

You need a copy of the sheet
each and a 5-minute timer.

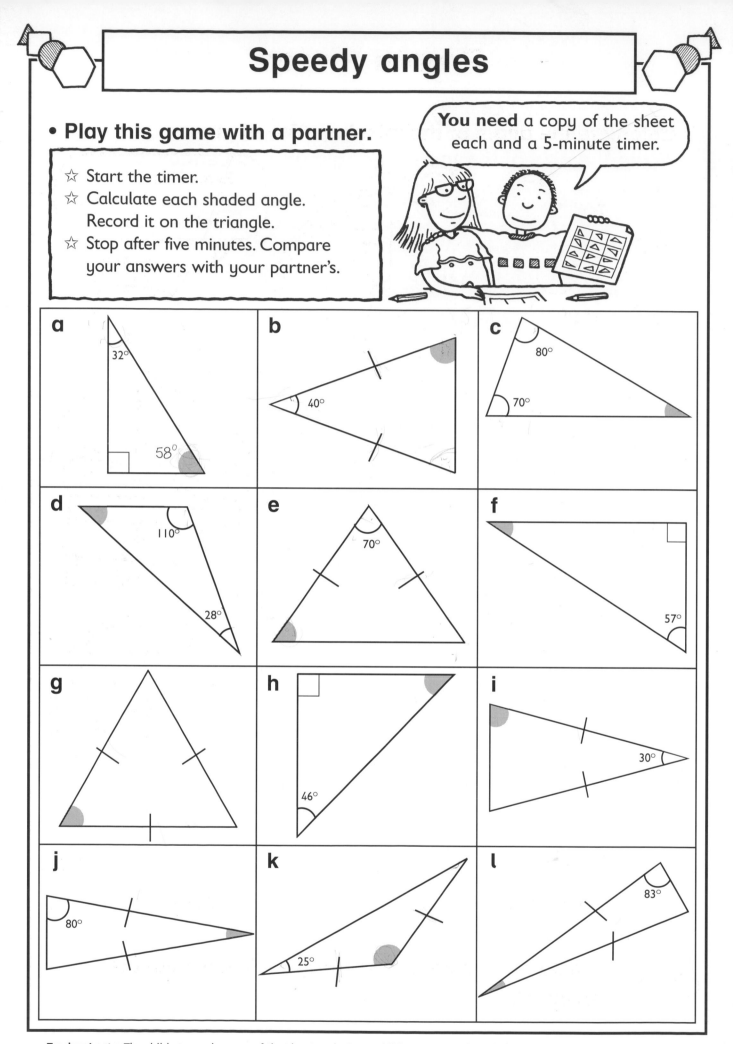

Teachers' note The children need a copy of the sheet each. Some children may need reminding about the angles of an isosceles and an equilateral triangle, and the meaning of the marks to show two sides are equal in length. The children can discuss any problems during the plenary session.

**Developing Numeracy
Measures, Shape and Space
Year 6**
© A & C Black

58

Pronto Pizzas

- **Calculate the angle of the missing slice of pizza.**

1. 70° 140° 150°

2. 45° 85°

Remember, angles which meet at a point add up to 360°.

3. 110° 90° 80°

4. 80° 85° 35°

5. 86° 142°

6. 135° 48° 140°

7. 65° 32° 73° 155°

- **Draw two** [intersecting] **straight lines. Measure one of the angles.**
- **Calculate the other three angles. Then measure to check.**

Teachers' note Challenge the children to calculate the missing angle mentally, then check using a pencil and paper. If necessary, introduce or revise the term 'intersecting'.

**Developing Numeracy
Measures, Shape and Space
Year 6
© A & C Black**

Spot the mistakes

Some of the angles on these triangles are marked incorrectly.

- Measure each angle. Mark the errors with a ☒. Write the correct angle.

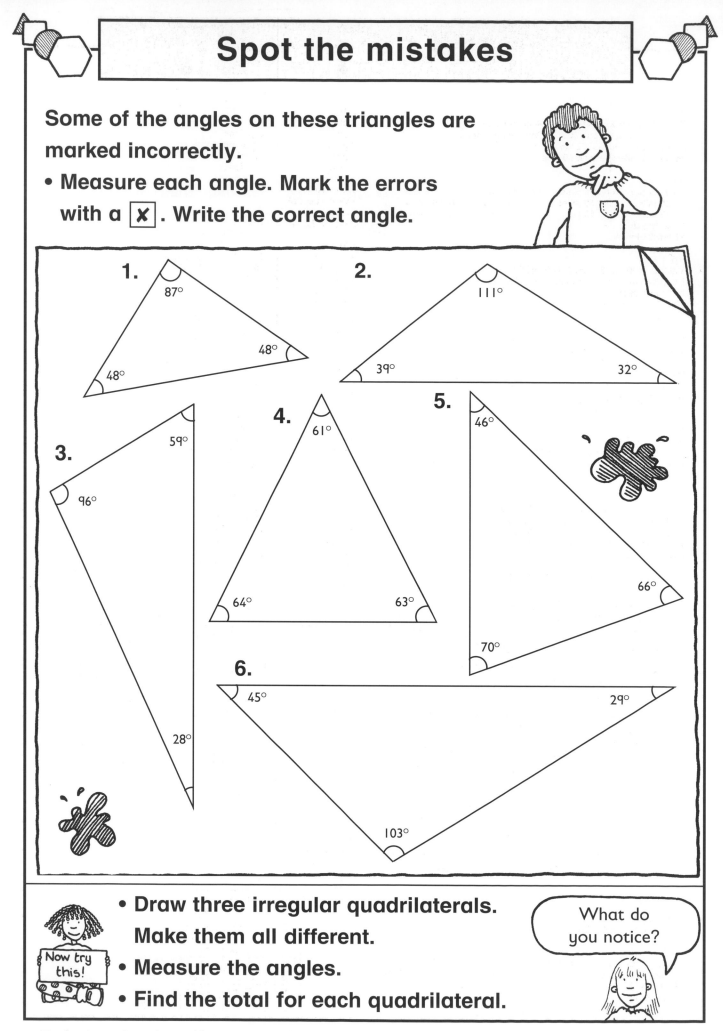

1.
87°
48°
48°

2.
111°
39°
32°

3.
59°
96°
28°

4.
61°
64°
63°

5.
46°
66°
70°

6.
45°
29°
103°

- **Draw three irregular quadrilaterals. Make them all different.**
- **Measure the angles.**
- **Find the total for each quadrilateral.**

Now try this!

What do you notice?

Teachers' note Stress the need for accurate positioning of the protractor, making sure that the centre is exactly on the vertex of the triangle, and that the zero line is exactly on the 'arm' of the angle to be measured.

Developing Numeracy
Measures, Shape and Space
Year 6
© A & C Black

p 6

Torland 24 km; Stirland 64 km; Oldquay 40 km
Portsea 5 miles; Porbay 10 miles; Westmouth 30 miles
Sunton 32 km; Bunter 72 km
Peasby 35 miles; Burland 50 miles

p 7

Answers are given to the nearest multiple of 5.

1. 25 miles	**2.** 60 miles	**3.** 85 miles	**4.** 40 miles
5. 95 km	**6.** 50 km	**7.** 80 km	**8.** 110 km

Now try this!
42 km; 21 km

p 8

3500 m; 1600 m; 2530 m; 31 000 m; 700 m;
3 km; 2·4 km; 1·75 km; 0·5 km; 10 km; 1·07 km

Now try this!
60 050 m; 60·05 km

p 9

a true	**b** false	**c** false	**d** true
e $13\frac{1}{2}$ ft; $4\frac{1}{2}$ yd	**f** 18 ft; 6 yd		
g $4\frac{1}{2}$ ft; $1\frac{1}{2}$ yd	**h** 39 ft; 13 yd		

Now try this!
horizontally: (78 × 4) + (21 × 2) = 354 ft
vertically: (36 × 3) + (27 × 2) = 162 ft
Total is 516 ft; 172 yd

p 10

Now try this!
5 cm; 50 cm; 500 mm; 600 mm; 800 mm; 1000 mm and 100 cm;
130 cm; 1500 mm; 2300 mm; 2500 mm; 260 cm; 3500 mm; 500 cm;
900 cm; 1000 cm

p 11

1. 1500 g	**2.** 5000 g	**3.** 2700 g
4. 6650 g	**5.** 3525 g	**6.** 550 g
7. 2 kg	**8.** 1·6 kg	**9.** 3·75 kg
10. 0·6 kg	**11.** 0·08 kg	**12.** 1·05 kg

Now try this!

1. £1·40	**2.** £3·50	**3.** £2·10
4. £4·90	**5.** £2·80	**6.** £0·70

Gold finds 1520 to 1530 (Charlie August)

Year	oz	g	Year	oz	g
1520	1 oz	30 g	1526	2 oz	60 g
1521	$1\frac{1}{2}$ oz	45 g	1527	3 oz	90 g
1522	$\frac{1}{2}$ oz	15 g	1528	$4\frac{1}{2}$ oz	135 g
1523	5 oz	150 g	1529	$3\frac{1}{2}$ oz	105 g
1524	4 oz	120 g	1530	$5\frac{1}{2}$ oz	165 g
1525	$2\frac{1}{2}$ oz	75 g	Total	33 oz	990 g

Gold finds 1520 to 1530 (Billy August)

Year	oz	g	Year	oz	g
1520	4 oz	120 g	1526	6 oz	180 g
1521	9 oz	270 g	1527	3 oz	90 g
1522	2 oz	60 g	1528	12 oz	360 g
1523	10 oz	300 g	1529	$1\frac{1}{4}$ lb	600 g
1524	5 oz	150 g	1530	2 lb	960 g
1525	1 oz	30 g	Total	104 oz	3120 g

Now try this!
Charlie £99   Billy £312

p 13

a 1 lb	**b** 4 lb	**c** 7 lb	**d** 12 lb	**e** 15 lb	**f** 19 lb
g 2 kg	**h** 6 kg	**i** 11 kg	**j** 13 kg	**k** 17 kg	**l** 23 kg

Now try this!
94 lb; 188 kg

p 15

2 pints	1 gallon
1 pint	2 gallons
7 pints	20 gallons
4 pints	10 gallons
22 pints	6 gallons
50 pints	4 gallons

Now try this!
VW Beetle: £24
Renault Clio: £34.40
Peugeot 206: 45 litres

p 16

1. 300 ml	**2.** 800 ml	**3.** 650 ml
4. 20 cl	**5.** 90 cl	**6.** 35 cl
7. 2·2 l	**8.** 0·32 l	**9.** 4·5 l
10. 0·035 l	**11.** 0·15 l	**12.** 0·75 l

Now try this!

1. 30 cl	**2.** 80 cl	**3.** 65 cl
4. 20 cl	**5.** 90 cl	**6.** 35 cl
7. 220 cl	**8.** 32 cl	**9.** 450 cl
10. 3·5 cl	**11.** 15 cl	**12.** 75 cl

p 18

Examples of relationships are:

pound	1 lb = 16 oz
foot	3 ft = 1 yd
millilitre	1000 ml = 1 l
centimetre	100 cm = 1 m
litre	1 l = 1000 ml
gallon	1 gallon = 8 pints
yard	1 yard = 36 inches
millimetre	10 mm = 1 cm
kilogram	1 kg = 1000 g
inch	12 inches = 1 foot
centilitre	100 cl = 1 l
gram	1000 g = 1 kg
ounce	16 oz = 1 lb
kilometre	1 km = 1000 m
pint	8 pints = 1 gallon
mile	5 miles is approximately 8 km

p 19

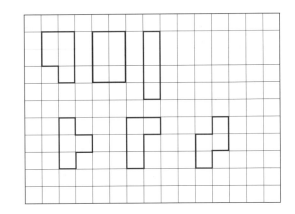

p 20

1. d, k, l   **2.** a, d, k
3. a, b, d, f, g, j, k, n

Now try this!
b, e, h, j, n

p 21

Shape	Area	Perimeter
a	38 cm²	34 cm
b	38 cm²	30 cm
c	32 cm²	34 cm
d	66 cm²	42 cm
e	45 cm²	36 cm
f	81 cm²	40 cm
g	48 cm²	38 cm
h	77 cm²	40 cm
i	56 cm²	36 cm

Now try this!
Perimeter: b; a and c; e and i; g; f and h; d
Area: c; a and b; e; g; i; d; h; f

p 22

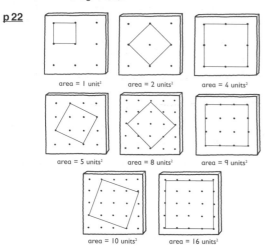

area = 1 unit²   area = 2 units²   area = 4 units²

area = 5 units²   area = 8 units²   area = 9 units²

area = 10 units²   area = 16 units²

The 5 units² and 10 units² oblique squares can be drawn as shown or reflected in the vertical mirror line.

p 23
1. 3500 yd²   **2.** 792 yd²   **3.** 120 yd²
4. 2708 yd²   **5.** 672 yd²   **6.** 6208 yd²
Now try this!
2720 yd; 11 laps

p 24
a 8 cm²   **b** 7·5 cm²   **c** 6 cm²   **d** 6 cm²   **e** 10·5 cm²
f 10 cm²   **g** 7 cm²   **h** 18 cm²   **i** 20 cm²   **j** 18 cm²

p 26
There are various possibilities.

p 27
1. 22 units²   **2.** 24 units²   **3.** 26 units²
4. 32 units²   **5.** 18 units²   **6.** 22 units²
7. 24 units²   **8.** 38 units²   **9.** 24 units²

p 28

Length of edge	Number of cubes needed	Surface area
1 cm	1	6 cm²
2 cm	8	24 cm²
3 cm	27	54 cm²
4 cm	64	96 cm²
5 cm	125	150 cm²
6 cm	216	216 cm²

Now try this!
10 cm; 1000; 600 cm²

p 29

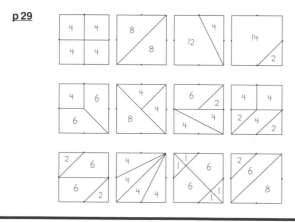

p 30

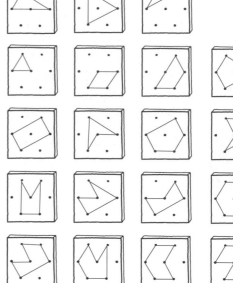

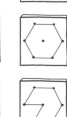

p 31
1. 63° and 117°   **2.** 76° and 104°
3. 78° and 102°   **4.** 65° and 115°
The opposite angles of a parallelogram are equal.

p 32

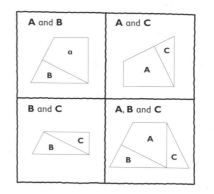

Now try this!

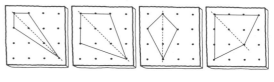

p 33

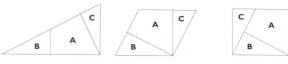

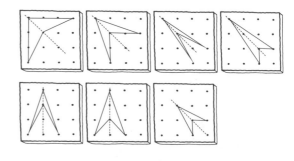

p 34

arrowhead — trapezium — square — parallelogram

trapezium — parallelogram — arrowhead

rectangle — kite — trapezium

square — square

p 35

quadrilaterals	rectangle	square	parallelogram	rhombus	trapezium	kite	arrowhead
4 sides	✓	✓	✓	✓	✓	✓	✓
all sides equal	?	✓	?	✓	✗	✗	✗
2 pairs of equal sides	✓	✓	✓	✓	✗	✓	✗
4 angles	✓	✓	✓	✓	✓	✓	✓
all angles equal	✓	✓	?	?	✗	✗	✗
2 pairs of equal angles	✓	✓	✓	✓	?	✗	✗
parallel sides	✓	✓	✓	✓	✓	✗	✗
perpendicular sides	✓	✓	?	?	?	?	?
at least 1 right angle	✓	✓	?	?	?	?	?
line symmetry	✓	✓	?	✓	?	✓	✓

p 37

Check diagonals are drawn correctly.

Quadrilateral	Same length?	Meet at mid-points?	Meet at right angles?
square	✓	✓	✓
rectangle	✓	✓	✗
rhombus	✗	✓	✓
parallelogram	✗	✓	✗
kite	✗	✗	✓
trapezium	✗	✗	✗

p 38

Possibilities include:

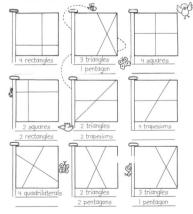

4 rectangles — 3 triangles 1 pentagon — 4 squares

2 squares 2 rectangles — 2 triangles 2 trapeziums — 4 trapeziums

4 quadrilaterals — 2 triangles 2 pentagons — 3 triangles 1 pentagon

p 39

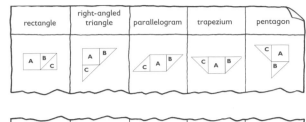

rectangle	right-angled triangle	parallelogram	trapezium	pentagon

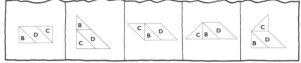

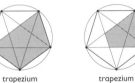

p 41

trapezium — trapezium — isosceles triangle

isosceles triangle — isosceles triangle — isosceles triangle

parallelogram — kite — arrowhead

Now try this!
5 triangles – the 4 shown plus:

p 43

Net	Shape name	Number of		
		faces	edges	vertices
a	cuboid	6	12	8
b	cube	6	12	8
c	tetrahedron*	4	6	4
d	triangular prism	5	9	6
e	square-based pyramid	5	8	5
f	pentagonal prism	7	15	10

* triangular-based pyramid

Now try this!

6 faces
10 edges
6 vertices

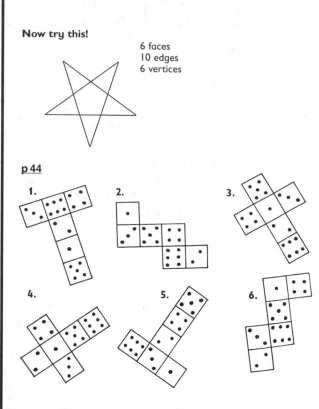

p 44

1. 2. 3. 4. 5. 6.

p 45

Net	Shape name	Number of		
		faces	edges	vertices
a	cube	6	12	8
b	hexagonal prism	8	18	12
c	tetrahedron*	4	6	4
d	pentagonal prism	7	15	10
e	triangular prism	5	9	6
f	square-based pyramid	5	8	5
g	octahedron	8	12	6
h	hexagonal-based pyramid	7	12	7

* triangular-based pyramid

The total of the number of faces and vertices is 2 more than the number of edges.

p 48
1. H   2. J   3. B   4. F   5. D
6. A   7. I   8. C   9. G   10. E

p 49
Quadrilaterals are as follows: kite, parallelogram, trapezium.
Check children's own quadrilaterals.

p 50

Quadrilateral	Lines of symmetry
parallelogram	0
square	4
rectangle	2
rhombus	2
trapezium	0
kite	1
arrowhead	1

Now try this!
The number of lines of symmetry of a regular polygon matches the number of sides.

p 51

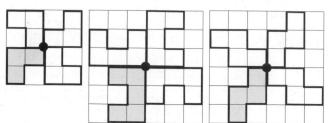

p 52
1. (⁻1, 1) (⁻1, 3) (1, 3) (1, 1)   2. (⁻1, 1) (⁻1, ⁻1) (⁻3, ⁻1) (⁻3, 1)
3. (1, ⁻2) (⁻1, ⁻2) (⁻1, ⁻4)   4. (1, ⁻2) (1, ⁻4) (3, ⁻4)
Now try this!
Shape A: (1, 1) (1, 3) (3, 1) (3, 3)
Shape C: (⁻2, ⁻1) (⁻2, ⁻3) (0, ⁻3)

p 53

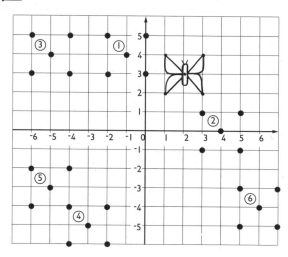

p 54
a 60°   b 72°   c 125°
d 108°   e 27°   f 66°

p 56
Check the children's charts.
The angles of a triangle add up to 180°.

p 57
1. 63°   2. 59°   3. 41°   4. 41°
5. 101°   6. 19°   7. 30°   8. 126°
9. 48°   10. 123°

p 58
a 58°   b 70°   c 30°   d 42°
e 55°   f 33°   g 60°   h 44°
i 75°   j 20°   k 130°   l 14°

p 59
1. 70°   2. 230°   3. 80°   4. 160°
5. 132°   6. 37°   7. 35°

p 60
1. 48° should be 45°   2. 111° should be 109°   3. 28° should be 25°
4. 61° should be 53°   5. 66° should be 64°   6. 29° should be 32°
Now try this!
The angles of a quadrilateral add up to 360°.